7/84

Dear Jan,

Here's a used copy. Sorry--

Evelyn

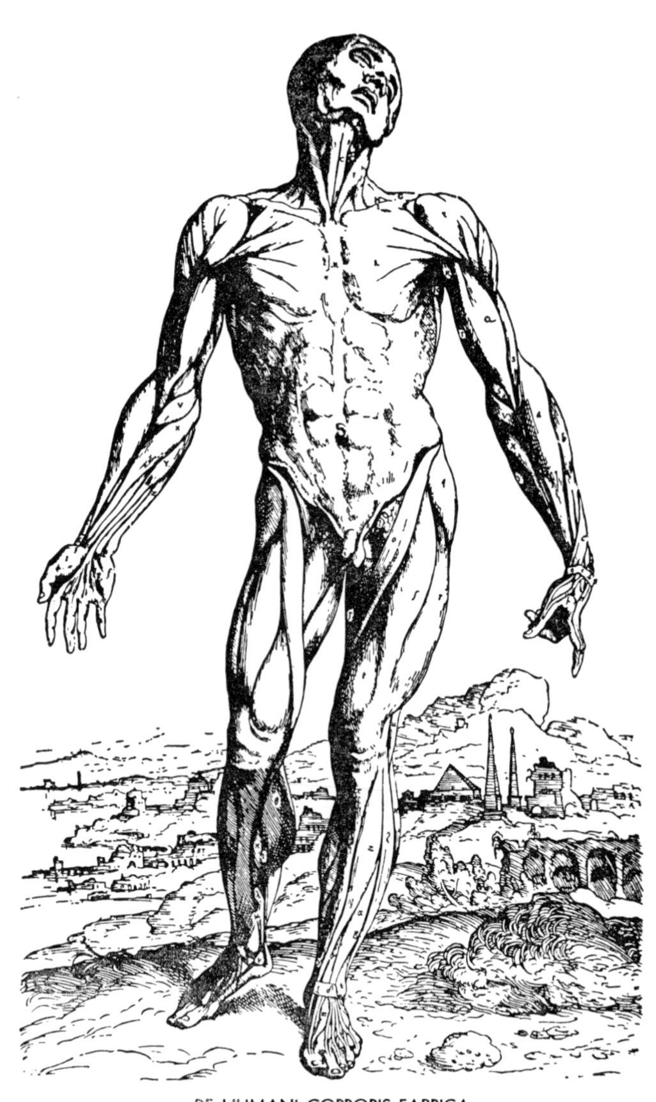

DE HUMANI CORPORIS FABRICA
ANDREAE VESALII
1568

THE EXTREMITIES

JOHN H. WARFEL, Ph.D.
ASSOCIATE PROFESSOR OF ANATOMY, STATE UNIVERSITY OF NEW YORK AT BUFFALO
SCHOOL OF MEDICINE, BUFFALO NEW YORK

Fourth Edition

110 Illustrations

LEA & FEBIGER

PHILADELPHIA

First Edition, 1945

Reprinted

June, 1946
April, 1948
September, 1950
January, 1952
October, 1953
November, 1954
September, 1956
February, 1958

Second Edition, 1960

Reprinted,

August, 1963

Third Edition, 1967

Reprinted

March, 1970

Fourth Edition, 1974

Reprinted

April, 1976
May, 1978
January, 1981
November, 1982

Warfel, John H.
The extremities.

First-3d editions by D. P. Quiring.
1. Extremities (Anatomy) 2. Muscles.
I. Quiring, Daniel Paul, 1894-1958. The extremities. II. Title. [DNLM: 1. Extremities—Anatomy and histology—Atlases. 2. Muscles—Anatomy and histology—Atlases. WE17 W275e 1974]
QM165.W27 1974 611'.737 73-7977
ISBN 0-8121-0457-9

ISBN 0-8121-0457-9

Library of Congress Catalog Card Number: 73:7977

Printed in the United States of America

PREFACE

In the first edition of The Extremities, authored by Daniel P. Quiring, late Head of the Anatomy Division of the Cleveland Clinic Foundation, the diagrams were designed to make clear to the student the origin, insertion, action and arterial and nerve supply of the muscles of the upper and lower extremities together with their motor points. Diagrammatic representation and condensed description, however, cannot do full justice to the complex relations involved. The plates do not attempt to show all details of attachments, nerves, and arteries, since the object is to emphasize the major termini of muscles and the chief arteries and nerves related to them. Likewise the legends stress the primary functions, which imply movement at the insertions. The student must realize, however, that when the insertions are fixed, muscles produce movement at their origins as well. In the lower extremity this occurs almost as frequently as primary action.

Motor points were tested on normal subjects. Since the points vary among individuals, diagrams can give only the approximate location of greatest muscular response. Motor points are not included for muscles which do not show a clear-cut response to electrical stimulus.

In this fourth edition reference corrections have been made to the 29th American Edition of Gray's Anatomy (Philadelphia, Lea & Febiger, 1973). The major change has been in the elimination of the references to Cunningham's Textbook of Anatomy. In their place references are now given to the 6th edition of Grant's Atlas of Anatomy (Baltimore, The Williams & Wilkins Co., 1972). The reader will notice that the references to Grant for the nerve and/or arterial supply to some of the muscles presented in this book are indicated as "not shown." This was done where it was found that the atlas did not have plates showing the specific relationship between the given muscle and its nerve or artery.

I feel that the reference combination of a leading textbook of anatomy with a popular atlas will enhance the service of this book to the user.

Buffalo, New York

John H. Warfel

CONTENTS

CONTENTS — Continued

DORSAL MUSCLES OF THE FOREARM

MUSCLES OF THE HAND

MUSCLES OF THE LOWER EXTREMITY, ILIAC REGION

ANTERIOR MUSCLES OF THE THIGH

CONTENTS — Continued

CONTENTS — Continued

TRAPEZIUS

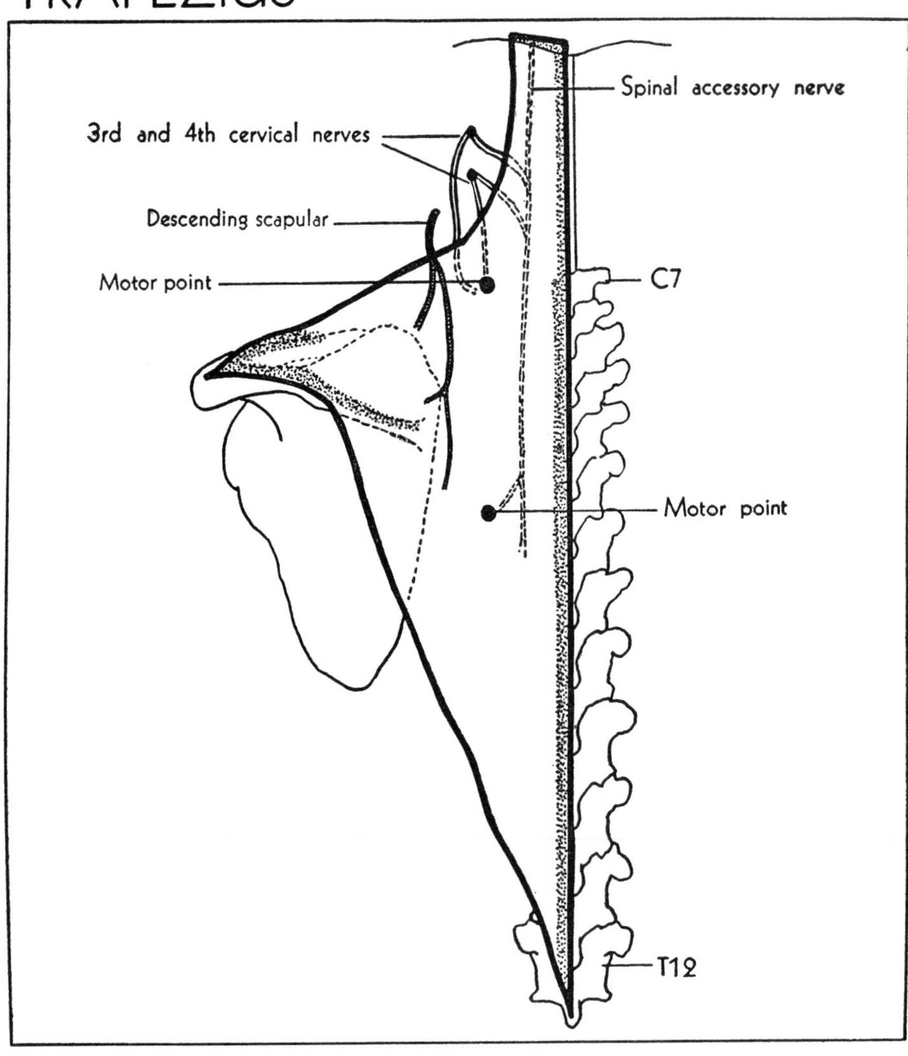

Spinal accessory nerve
3rd and 4th cervical nerves
Descending scapular
Motor point
C7
Motor point
T12

ORIGIN: External occipital protuberance, superior nuchal line, nuchal ligament from spines of seventh cervical and all thoracic vertebrae

INSERTION: Lateral third of clavicle, spine of scapula, acromion

FUNCTION: Adducts scapula, tilts chin, draws back acromion, rotates scapula

NERVE: Spinal accessory, 3d and 4th cervical

ARTERY: Descending scapular (transverse cervical)

REFERENCES:	GRAY	GRANT'S ATLAS
Muscle	447	24, 473
Nerve	448, 945, 955-956, 960	24, 25, 472, 661
Artery	606	25

LATISSIMUS DORSI

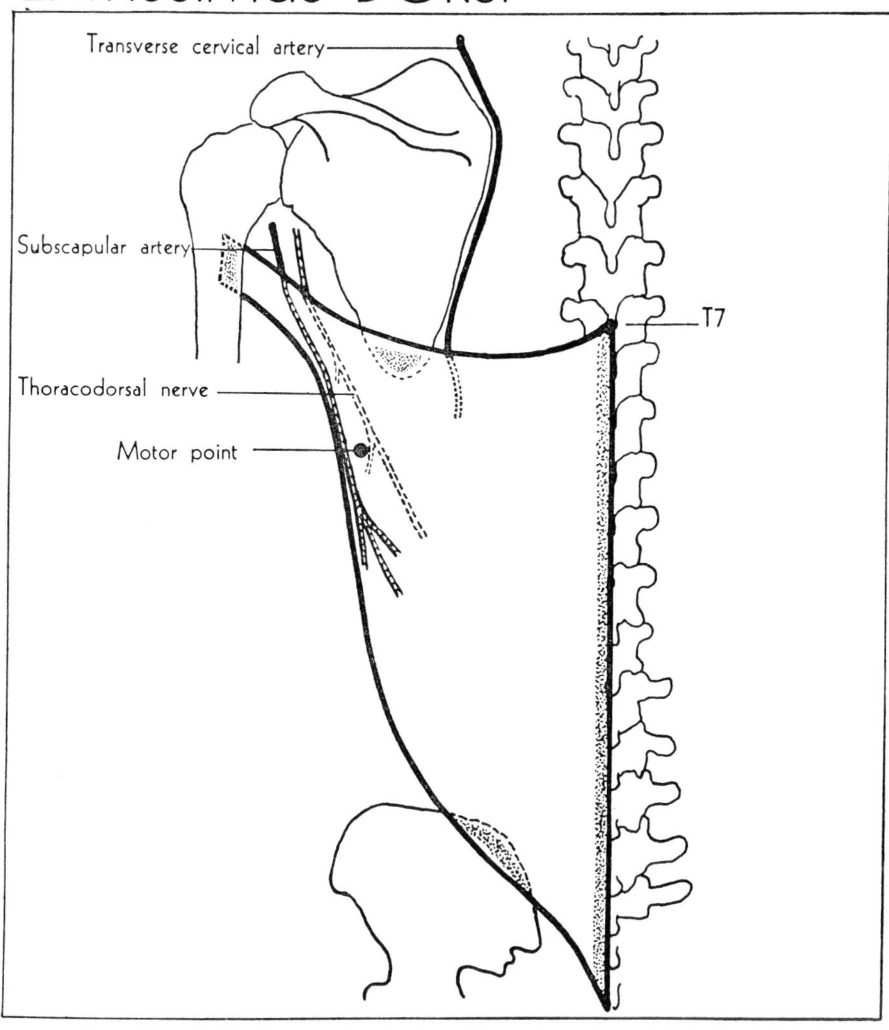

Transverse cervical artery
Subscapular artery
Thoracodorsal nerve
Motor point
T7

ORIGIN: Spines of lower 6 thoracic vertebrae, lumbodorsal fascia, crest of ilium, muscular slips from lower 3 or 4 ribs

INSERTION: Floor of bicipital groove of humerus

FUNCTION: Adducts, extends, and medially rotates humerus

NERVE: Thoracodorsal

ARTERY: Descending scapular, (transverse cervical), subscapular

REFERENCES: GRAY		GRANT'S ATLAS
Muscle	448	24
Nerve	448, 961, 963, 964	15, 17, 18
Artery	606, 613	17

RHOMBOIDEUS MAJOR

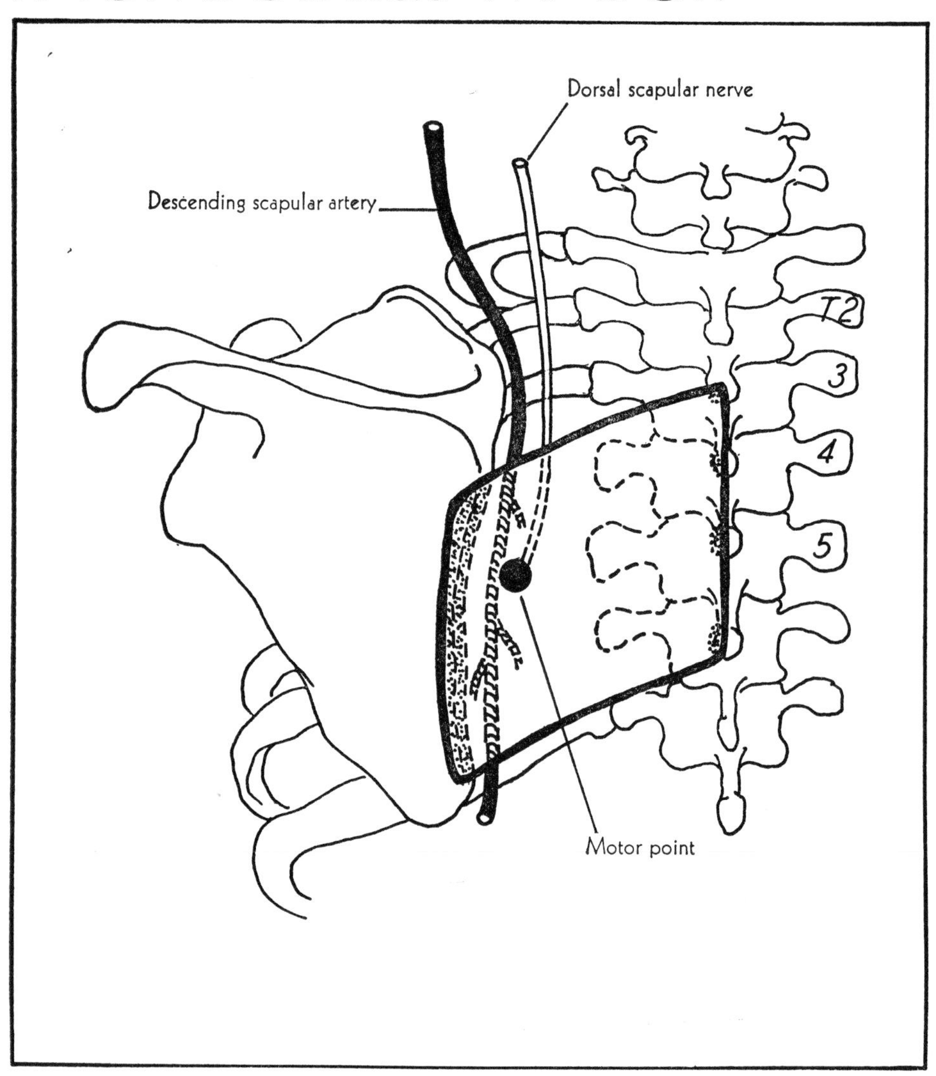
Dorsal scapular nerve
Descending scapular artery
T2
3
4
5
Motor point

ORIGIN:	Spine of 2d, 3d, 4th and 5th thoracic vertebra
INSERTION:	Medial border of scapula, between spine and inferior angle
FUNCTION:	Adducts and rotates scapula
NERVE:	Dorsal scapular
ARTERY:	Descending scapular (transverse cervical)

REFERENCES:	GRAY	GRANT'S ATLAS
Muscle	449	24, 478
Nerve	449, 961, 963	473, 474
Artery	606	Not shown

RHOMBOIDEUS MINOR

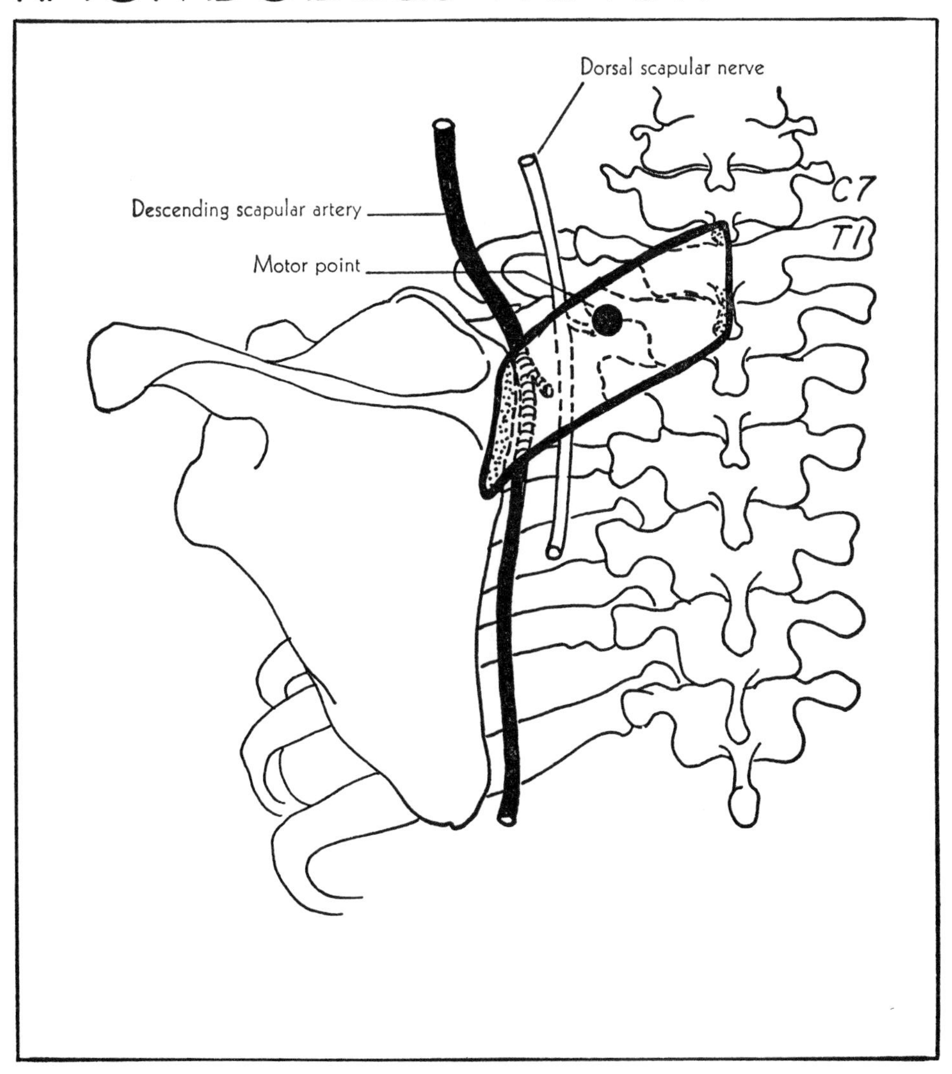

Dorsal scapular nerve
C7
T1
Descending scapular artery
Motor point

ORIGIN: Ligamentum nuchae, spine of 7th cervical and 1st thoracic vertebra
INSERTION: Root of scapular spine
FUNCTION: Adducts and rotates scapula
NERVE: Dorsal scapular
ARTERY: Descending scapular (transverse cervical)

REFERENCES:	GRAY	GRANT'S ATLAS
Muscle	449	24, 478
Nerve	449, 961, 963	473, 474
Artery	606	Not shown

LEVATOR SCAPULAE

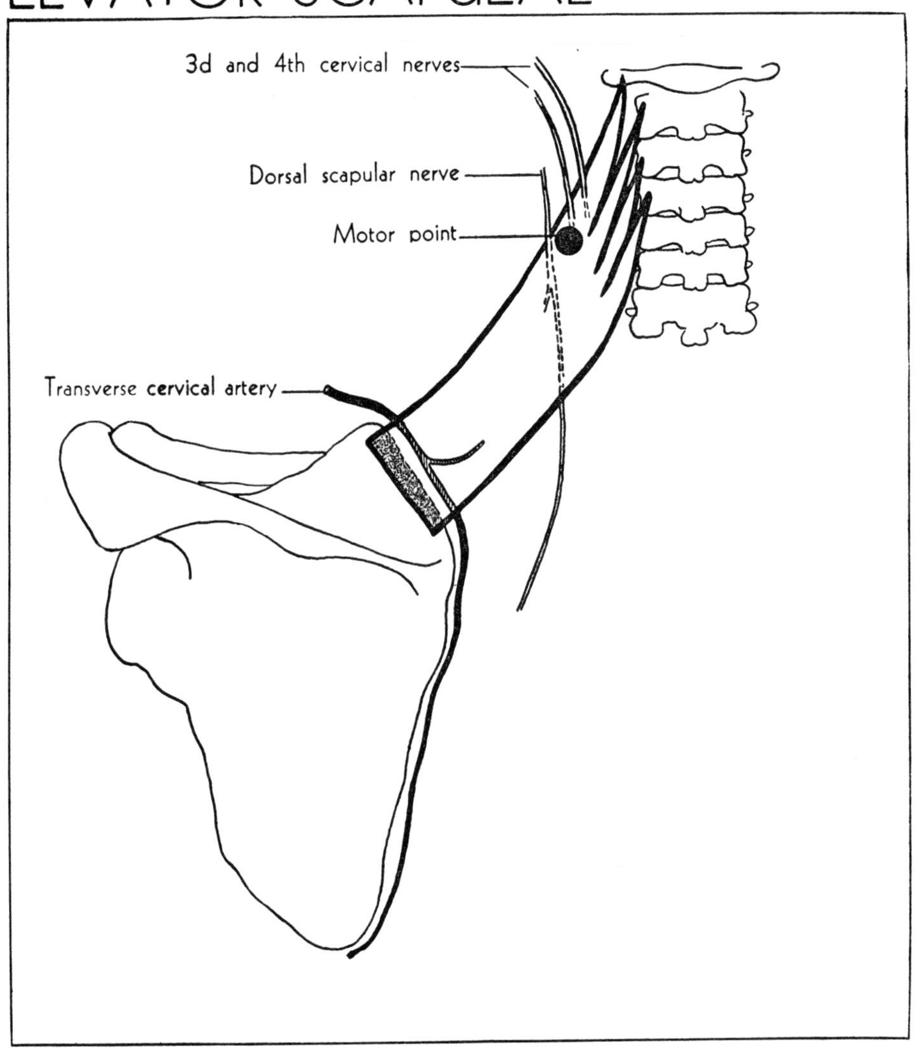
3d and 4th cervical nerves
Dorsal scapular nerve
Motor point
Transverse cervical artery

ORIGIN:	Transverse process of atlas, axis, 3d and 4th cervical vertebra
INSERTION:	Vertebral border of scapula between medial angle and root of spine
FUNCTION:	Raises scapula or inclines neck to corresponding side if scapula is fixed
NERVE:	Dorsal scapular, 3rd and 4th cervical
ARTERY:	Descending scapular (transverse cervical)

REFERENCES:	GRAY	GRANT'S ATLAS
Muscle	449	24, 478
Nerve	449, 955, 961, 963	473, 474
Artery	606	474, 475

PECTORALIS MAJOR

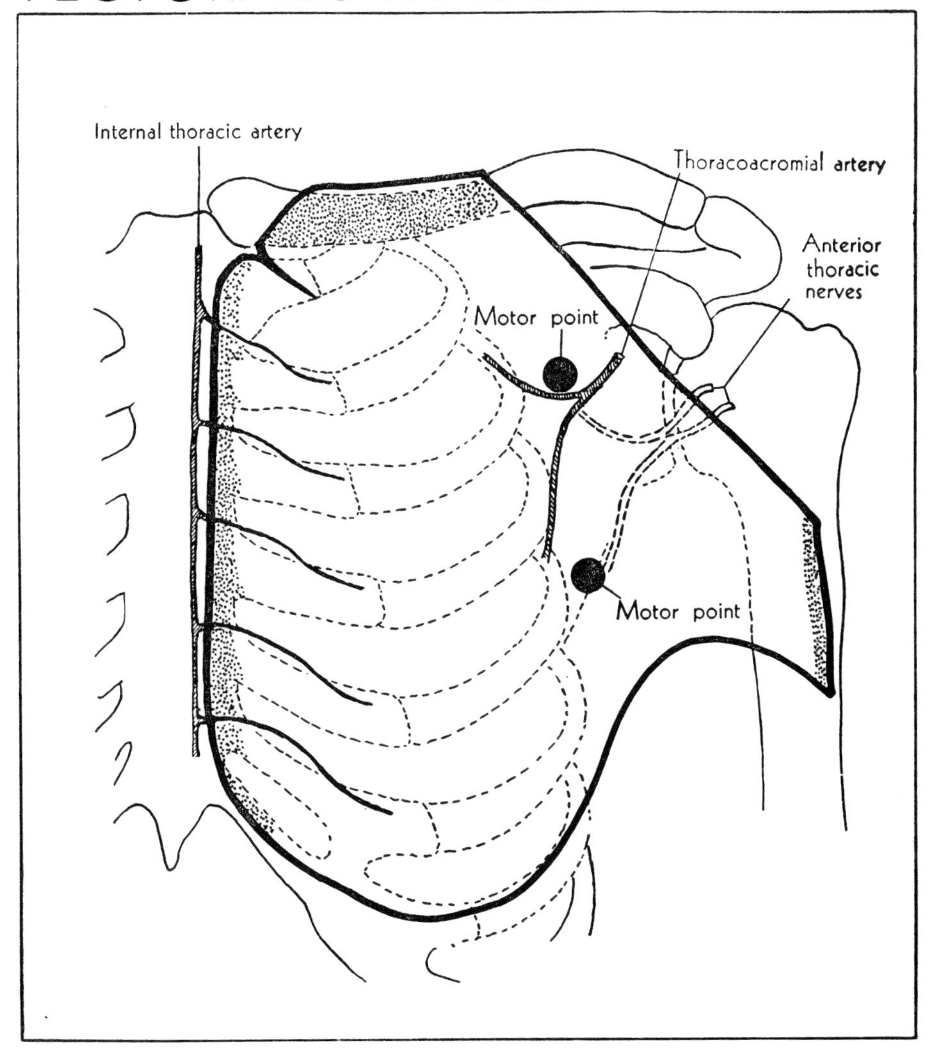
Internal thoracic artery
Thoracoacromial artery
Anterior thoracic nerves
Motor point
Motor point

ORIGIN: Sternal half of clavicle, sternum to 7th rib, cartilages of true ribs, aponeurosis of external oblique muscle

INSERTION: Lateral lip of bicipital groove of humerus

FUNCTION: Adducts arm, draws it forward, rotates it medially

NERVE: Medial and lateral anterior thoracic (med. & lat. pectoral nn.)

ARTERY: Pectoral branch of thoracoacromial, perforating branches of internal thoracic

REFERENCES:	GRAY	GRANT'S ATLAS
Muscle	453	13
Nerve	453, 961, 963, 964	14, 16
Artery	609, 612	14, 16

PECTORALIS MINOR

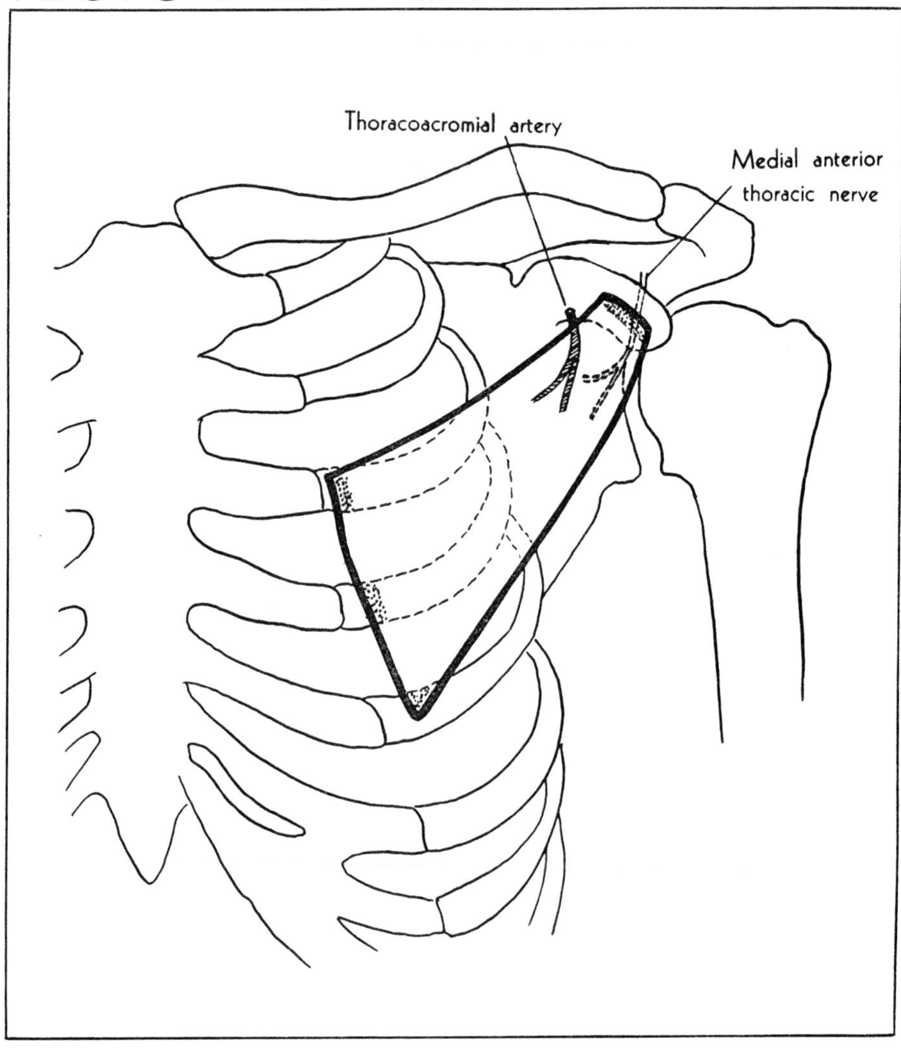
Thoracoacromial artery
Medial anterior thoracic nerve

ORIGIN: Outer surface of upper margin of 3d, 4th, and 5th rib
INSERTION: Coracoid process of scapula
FUNCTION: Lowers lateral angle of scapula; pulls shoulder forward
NERVE: Medial anterior thoracic (med. pectoral n.)
ARTERY: Thoracoacromial and intercostal branches of internal thoracic; lateral thoracic

REFERENCES:	GRAY	GRANT'S ATLAS
Muscle	454	406
Nerve	454, 961, 963, 964	16
Artery	609, 612, 613	16

SUBCLAVIUS

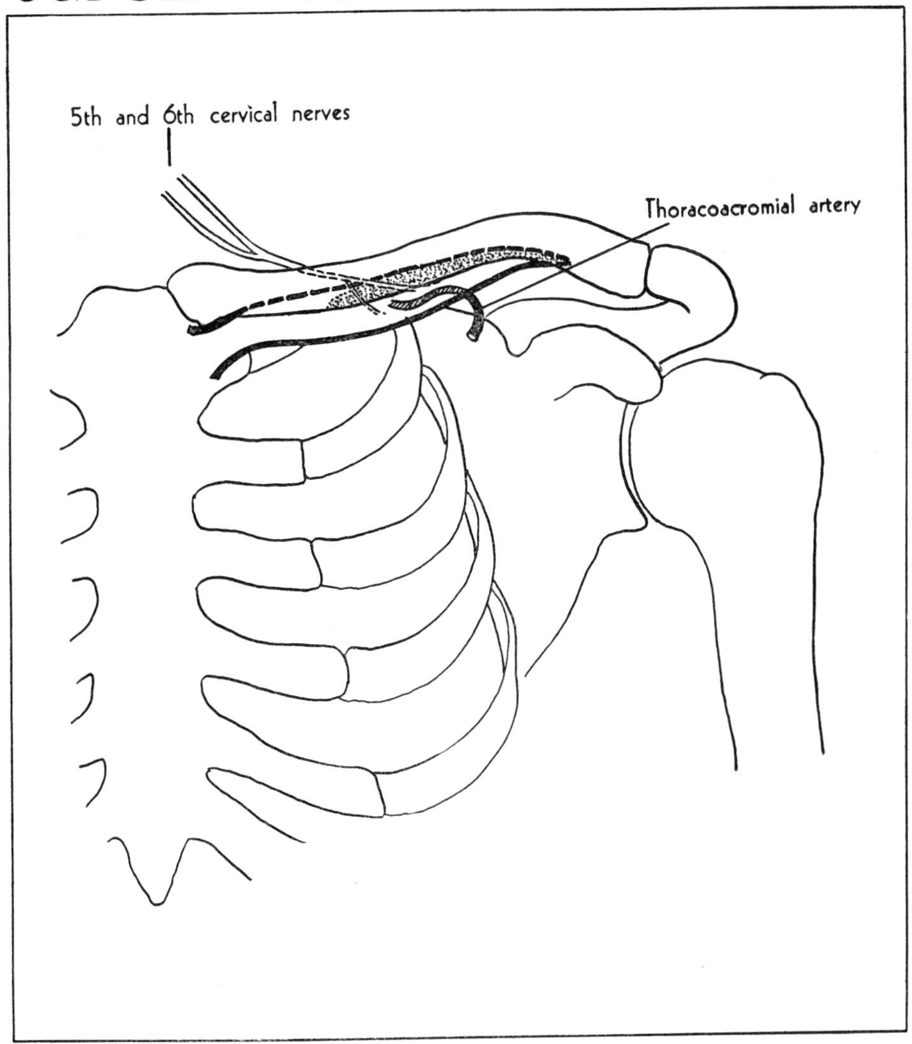

5th and 6th cervical nerves
Thoracoacromial artery

ORIGIN:	Upper border of 1st rib and its cartilage
INSERTION:	Groove on under surface of clavicle
FUNCTION:	Draws clavicle down and forward
NERVE:	5th and 6th cervical (Nerve to subclavius)
ARTERY:	Clavicular branch of thoracoacromial

REFERENCES:	GRAY	GRANT'S ATLAS
Muscle	454	16, 475
Nerve	454, 961, 963	475
Artery	613	14

SERRATUS ANTERIOR

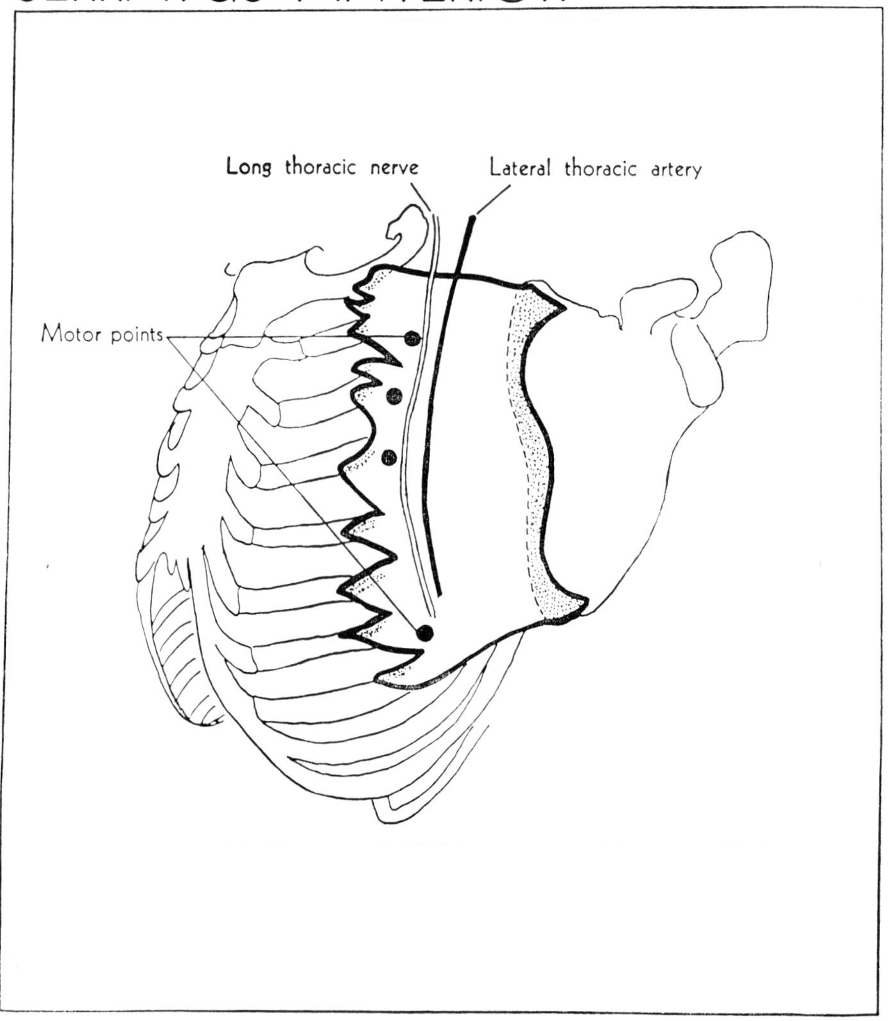
Long thoracic nerve
Lateral thoracic artery
Motor points

ORIGIN: Outer surface of upper 8 or 9 ribs
INSERTION: Costal surface of vertebral border of scapula
FUNCTION: Abducts scapula; raises ribs when scapula is fixed
NERVE: Long thoracic
ARTERY: Lateral thoracic

REFERENCES:	GRAY	GRANT'S ATLAS
Muscle	454	23, 478
Nerve	454, 961, 963	18, 23, 107
Artery	613	16

DELTOIDEUS

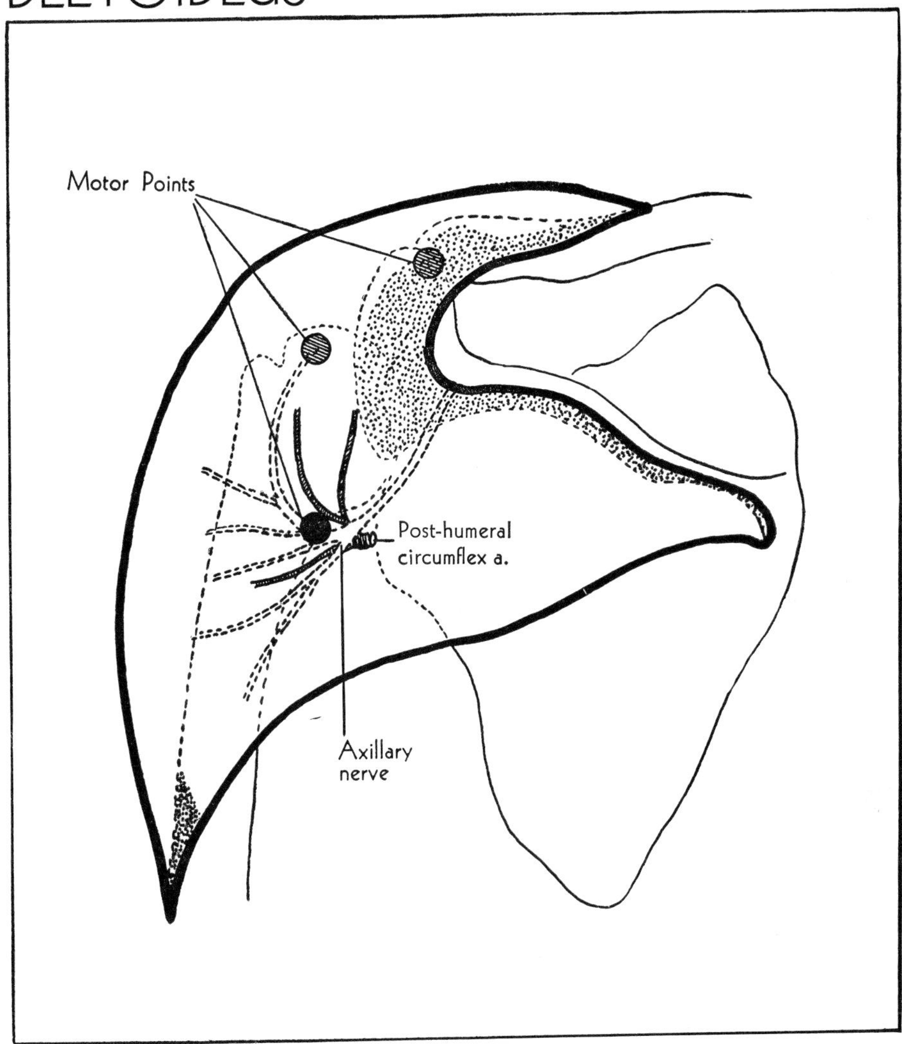

Motor Points
Post-humeral
circumflex a.
Axillary
nerve

ORIGIN: Lateral third of clavicle, upper surface of acromion, spine of scapula
INSERTION: Deltoid tuberosity of humerus
FUNCTION: Abducts arm
NERVE: Anterior and posterior branches of axillary (circumflex)
ARTERY: Posterior humeral circumflex; deltoid branch of thoracoacromial

REFERENCES:	GRAY	GRANT'S ATLAS
Muscle	455	14, 31
Nerve	455, 961, 963, 964	19, 32
Artery	613, 614	32

SUBSCAPULARIS

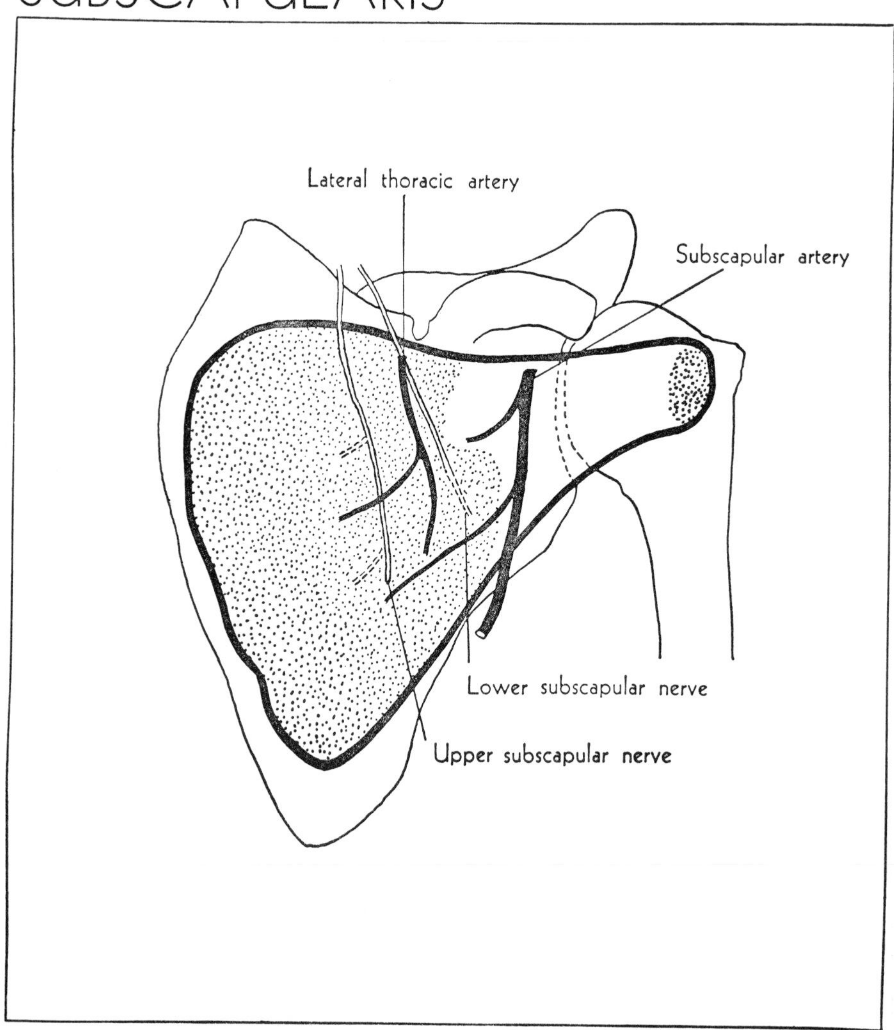
Lateral thoracic artery
Subscapular artery
Lower subscapular nerve
Upper subscapular nerve

ORIGIN:	Subscapular fossa
INSERTION:	Lesser tuberosity of humerus and capsule of shoulder joint
FUNCTION:	Rotates humerus medially, draws it forward and down when arm is raised
NERVE:	Upper and lower subscapular
ARTERY:	Lateral thoracic, subscapular

REFERENCES: GRAY		GRANT'S ATLAS
Muscle	455	18, 39
Nerve	456, 961, 963, 964	15, 18
Artery	613	16, 17

SUPRASPINATUS

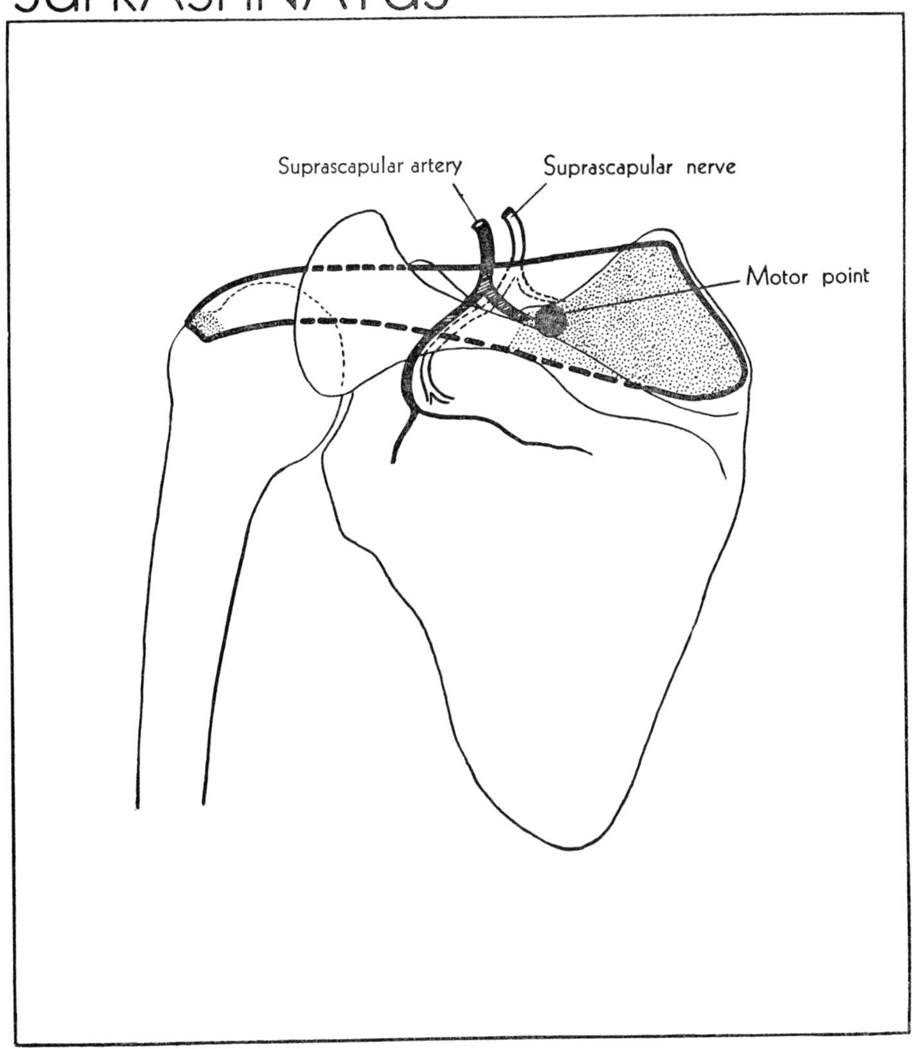

Suprascapular artery
Suprascapular nerve
Motor point

ORIGIN: Supraspinous fossa of scapula

INSERTION: Superior facet of greater tuberosity of humerus; capsule of shoulder joint

FUNCTION: Assists deltoid in abducting arm, fixes head of humerus in glenoid cavity; rotates head of humerus laterally

NERVE: Suprascapular

ARTERY: Suprascapular (transverse scapular)

REFERENCES:	GRAY	GRANT'S ATLAS
Muscle	456	26, 39
Nerve	456, 961, 963, 964	25, 32
Artery	605	25

INFRASPINATUS

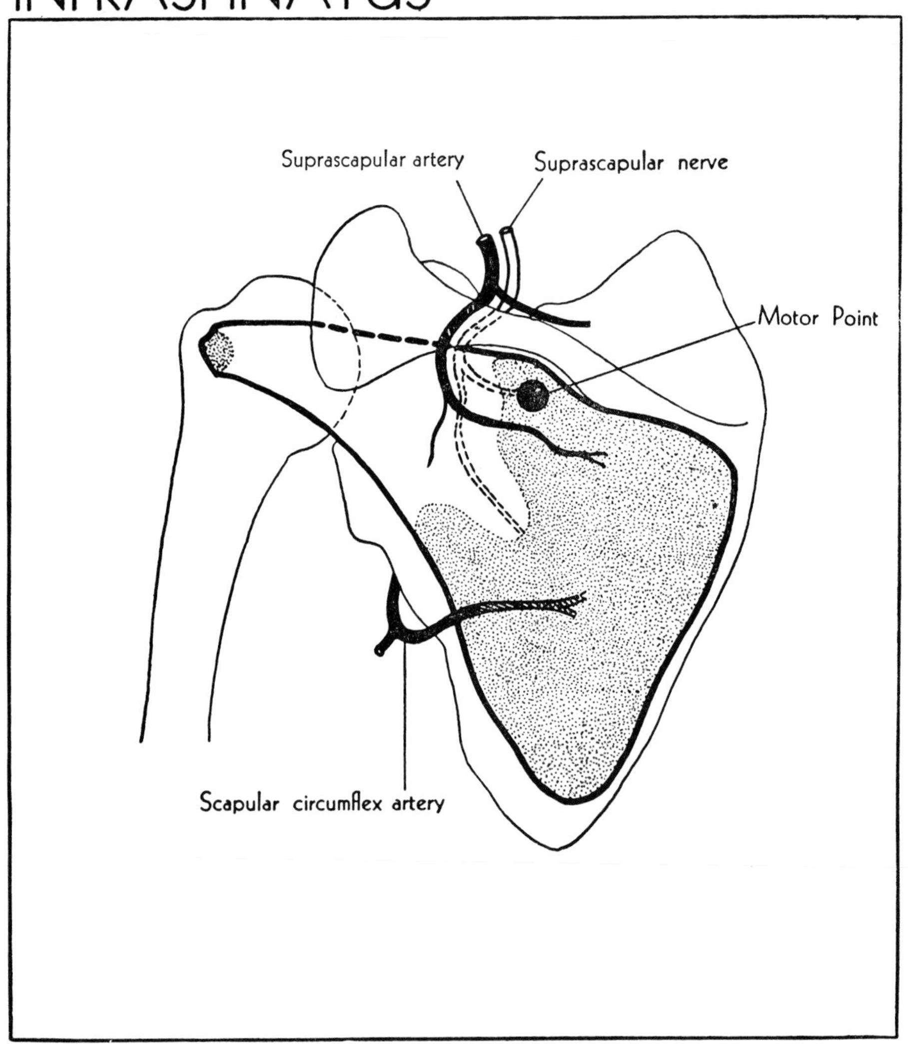

Suprascapular artery
Suprascapular nerve
Motor Point
Scapular circumflex artery

ORIGIN: Infraspinous fossa of scapula
INSERTION: Middle facet of greater tuberosity of humerus; capsule of shoulder joint
FUNCTION: Rotates head of humerus laterally with teres minor
NERVE: Suprascapular
ARTERY: Suprascapular (transverse scapular); scapular circumflex

REFERENCES:	GRAY	GRANT'S ATLAS
Muscle	456	32, 39
Nerve	457, 961, 963, 964	25, 32
Artery	605, 613	32

TERES MINOR

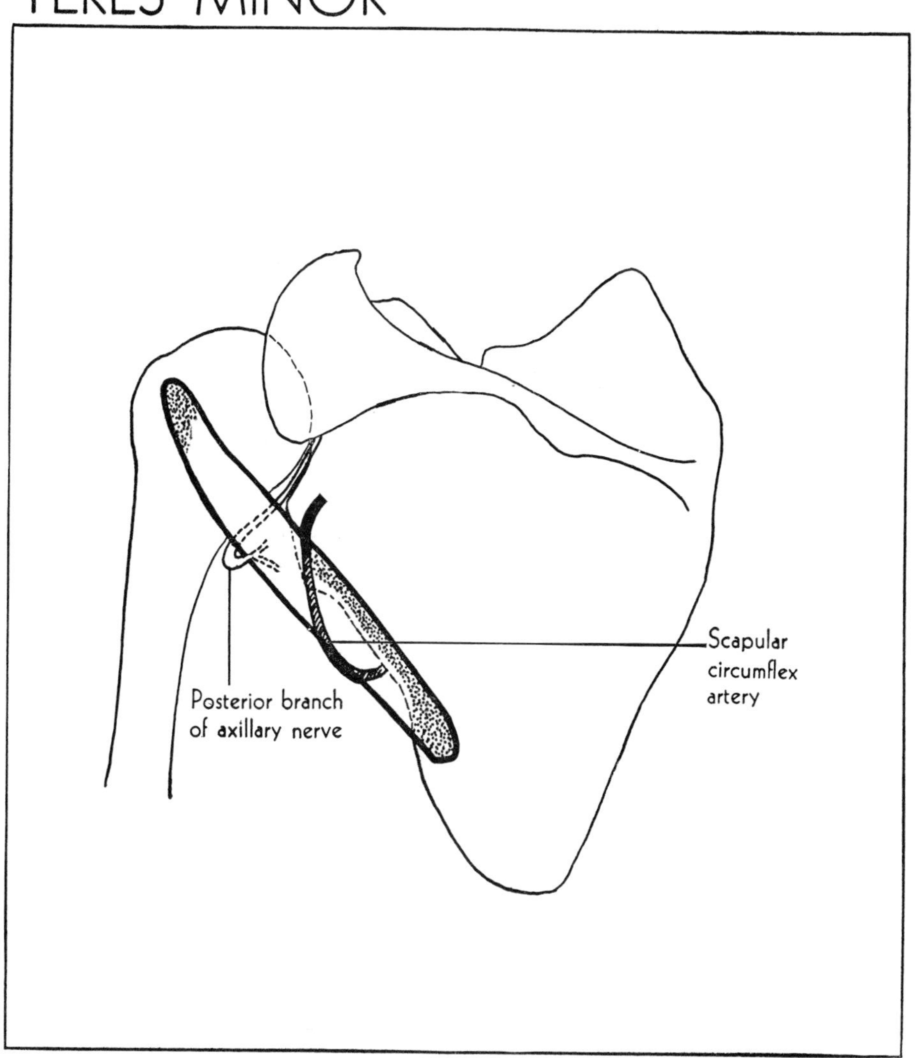

Scapular
circumflex
artery
Posterior branch
of axillary nerve

ORIGIN: Dorsal surface of axillary border of scapula

INSERTION: Lowest facet of greater tuberosity of humerus; capsule of shoulder joint

FUNCTION: Adducts and rotates head of humerus laterally and draws humerus toward glenoid fossa

NERVE: Posterior branch of axillary (circumflex)

ARTERY: Scapular circumflex

REFERENCES:	GRAY	GRANT'S ATLAS
Muscle	457	32, 39
Nerve	457, 961, 963, 964	19, 32
Artery	613	32

TERES MAJOR

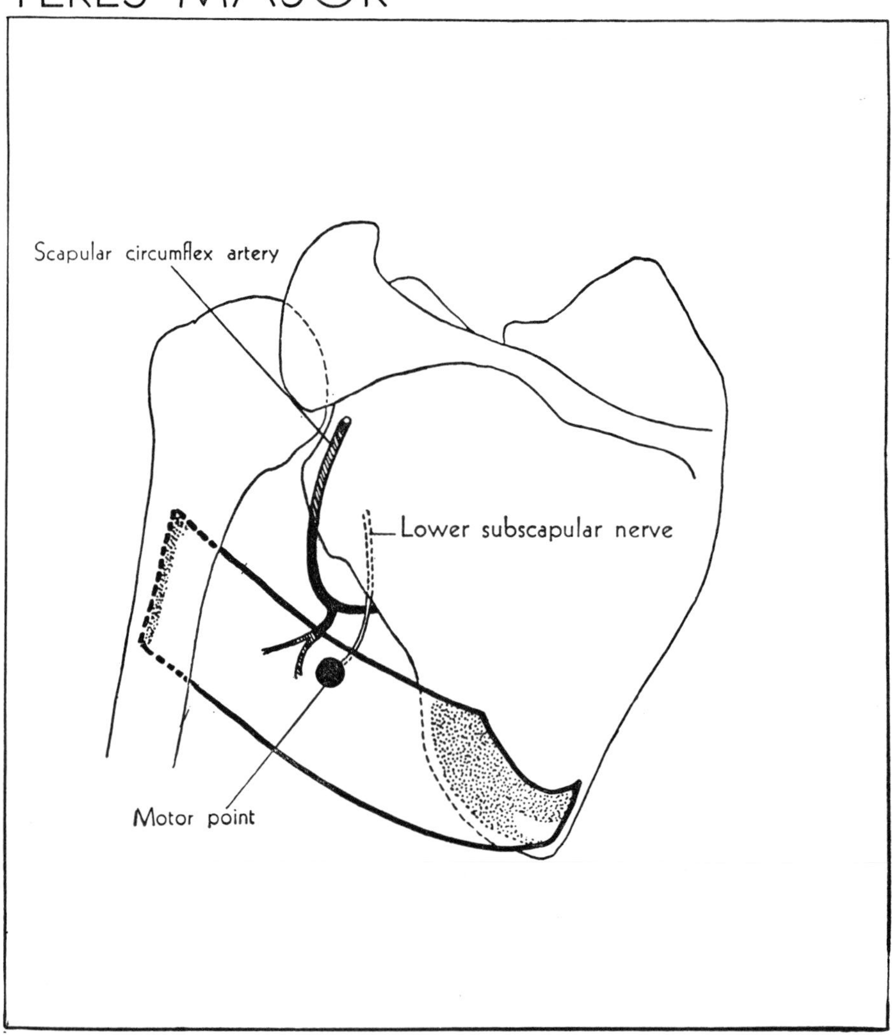
Scapular circumflex artery
Lower subscapular nerve
Motor point

ORIGIN: Dorsal surface of inferior angle of scapula
INSERTION: Medial lip of bicipital groove of humerus
FUNCTION: Adducts and medially rotates humerus and draws it back
NERVE: Lower subscapular
ARTERY: Scapular circumflex

REFERENCES:	GRAY	GRANT'S ATLAS
Muscle	457	18, 32
Nerve	457, 961, 963, 964	15, 18
Artery	613	32

CORACOBRACHIALIS

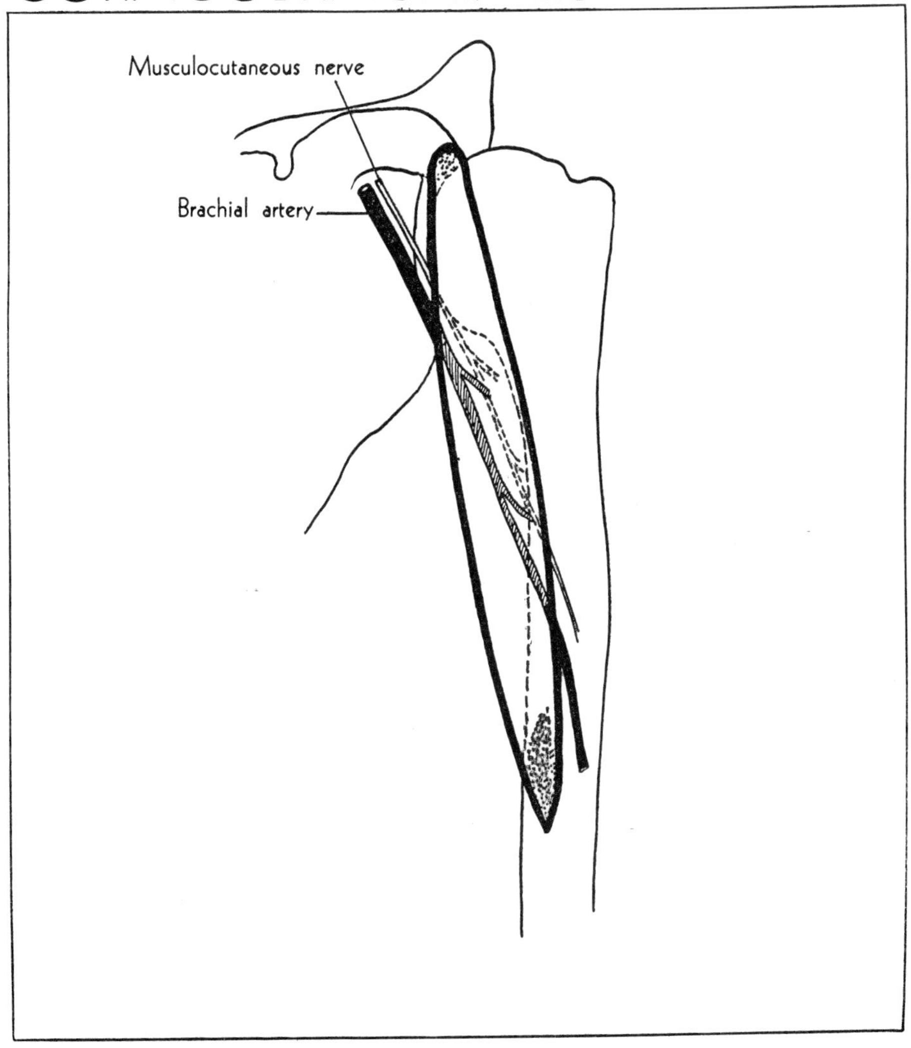
Musculocutaneous nerve
Brachial artery

ORIGIN: Tip of coracoid process of scapula
INSERTION: Middle of medial border of humerus
FUNCTION: Flexion and adduction of arm
NERVE: Musculocutaneous
ARTERY: Muscular branches of brachial

REFERENCES:	GRAY	GRANT'S ATLAS
Muscle	459	16, 18
Nerve	459, 961, 963, 969	18-20
Artery	617	20

BICEPS BRACHII

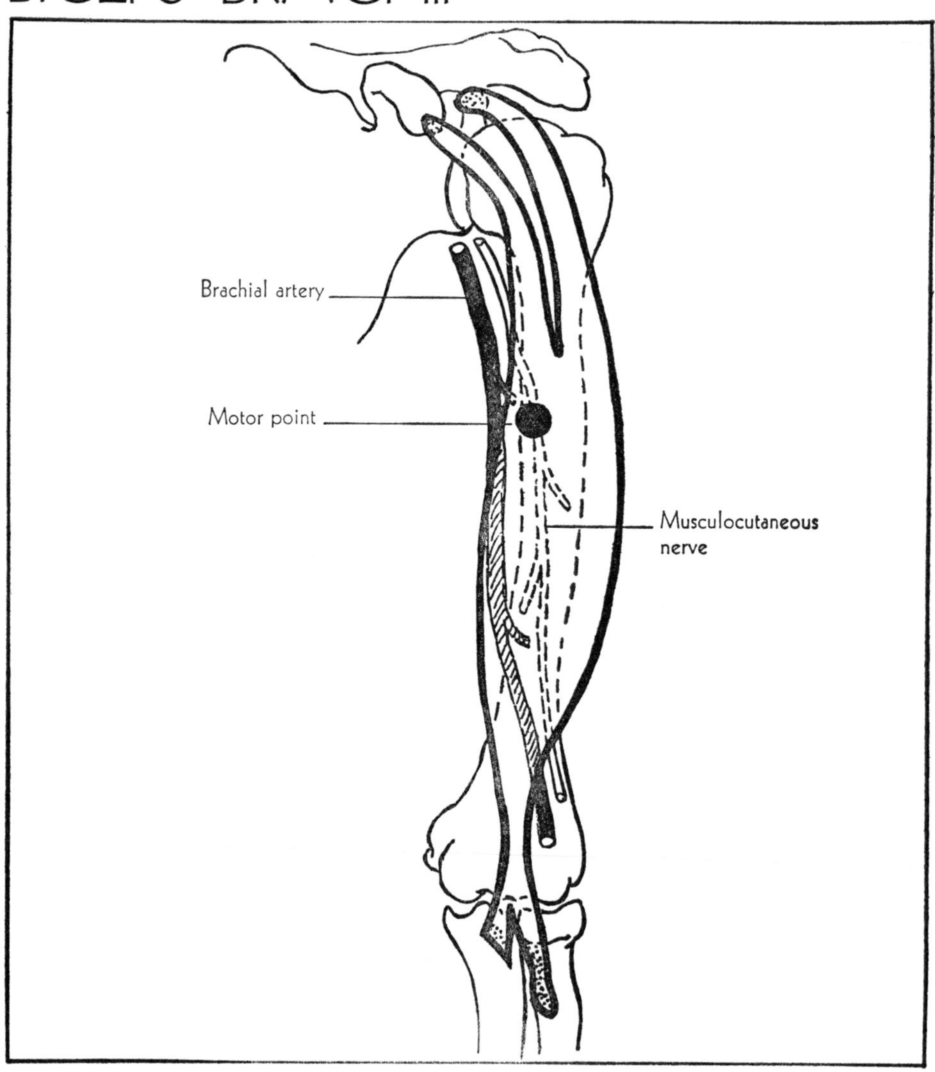

Brachial artery
Motor point
Musculocutaneous nerve

ORIGIN:	Short head from tip of coracoid process of scapula, long head from supraglenoid tuberosity of scapula
INSERTION:	Radial tuberosity and by lacertus fibrosus to origins of forearm flexors
FUNCTION:	Flexes and supinates forearm, flexes arm when forearm is fixed
NERVE:	Musculocutaneous
ARTERY:	Muscular branches of brachial

REFERENCES:	GRAY	GRANT'S ATLAS
Muscle	459	15, 20, 36, 45
Nerve	460, 961, 963, 969	18
Artery	617	18, 20

BRACHIALIS

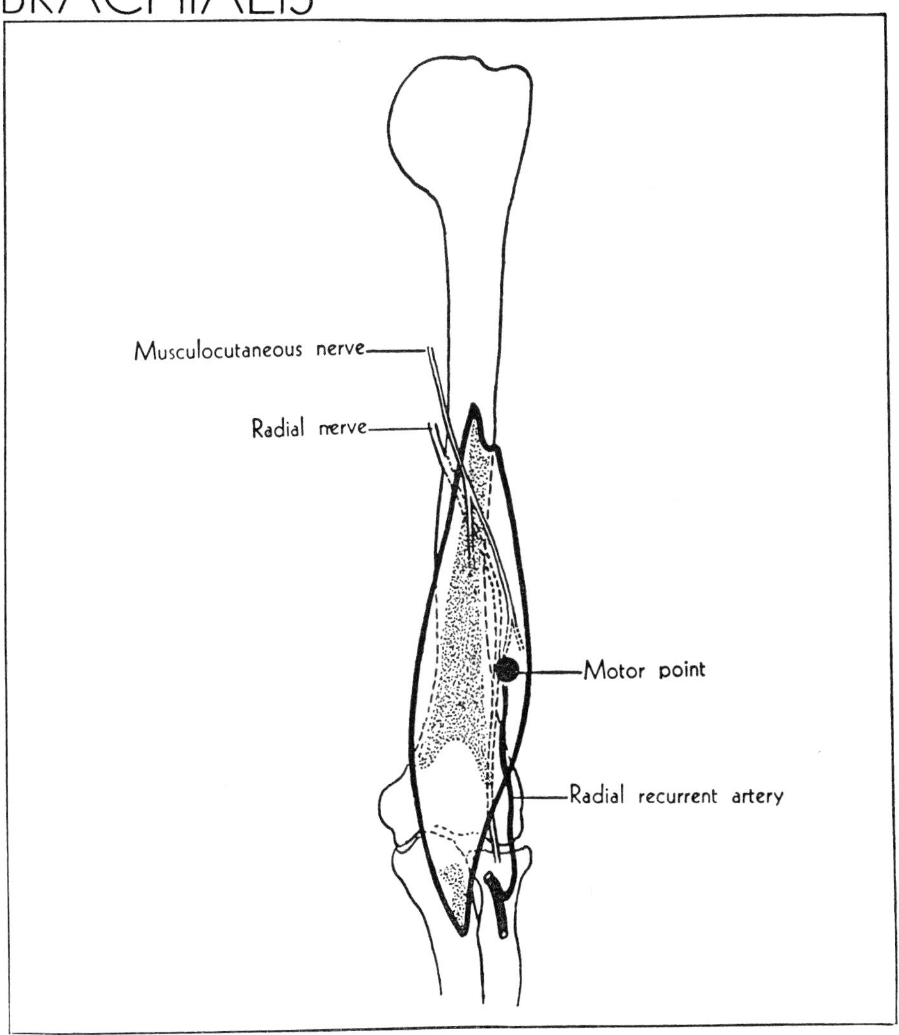

Musculocutaneous nerve
Radial nerve
Motor point
Radial recurrent artery

ORIGIN: Lower two-thirds of anterior surface of humerus
INSERTION: Coronoid process and tuberosity of ulna
FUNCTION: Flexes forearm
NERVE: Musculocutaneous, radial (may be afferent)
ARTERY: Radial recurrent, brachial

REFERENCES:	GRAY	GRANT'S ATLAS
Muscle	460	31, 46
Nerve	460, 961, 963, 969, 978	18
Artery	617, 620	20

TRICEPS BRACHII

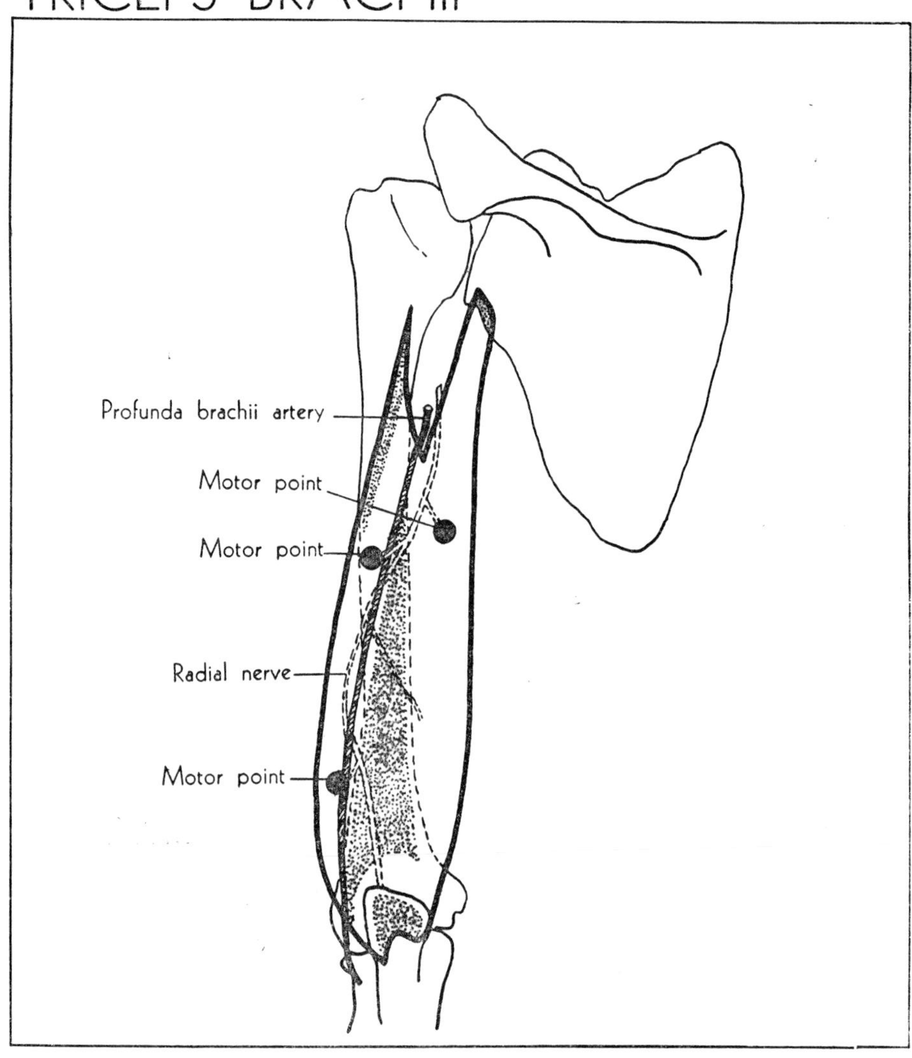

Profunda brachii artery
Motor point
Motor point
Radial nerve
Motor point

ORIGIN:	Long head from infraglenoid tuberosity of scapula, lateral head from posterior and lateral surface of humerus, medial head from lower posterior surface of humerus
INSERTION:	Upper posterior surface of olecranon and deep fascia of forearm
FUNCTION:	Extends forearm; if arm is abducted, long head aids in adducting it
NERVE:	Radial
ARTERY:	Branch of profunda brachii

REFERENCES:	GRAY	GRANT'S ATLAS
Muscle	460	20, 32, 56
Nerve	461, 961, 963, 976	33
Artery	616	18, 33

PRONATOR TERES

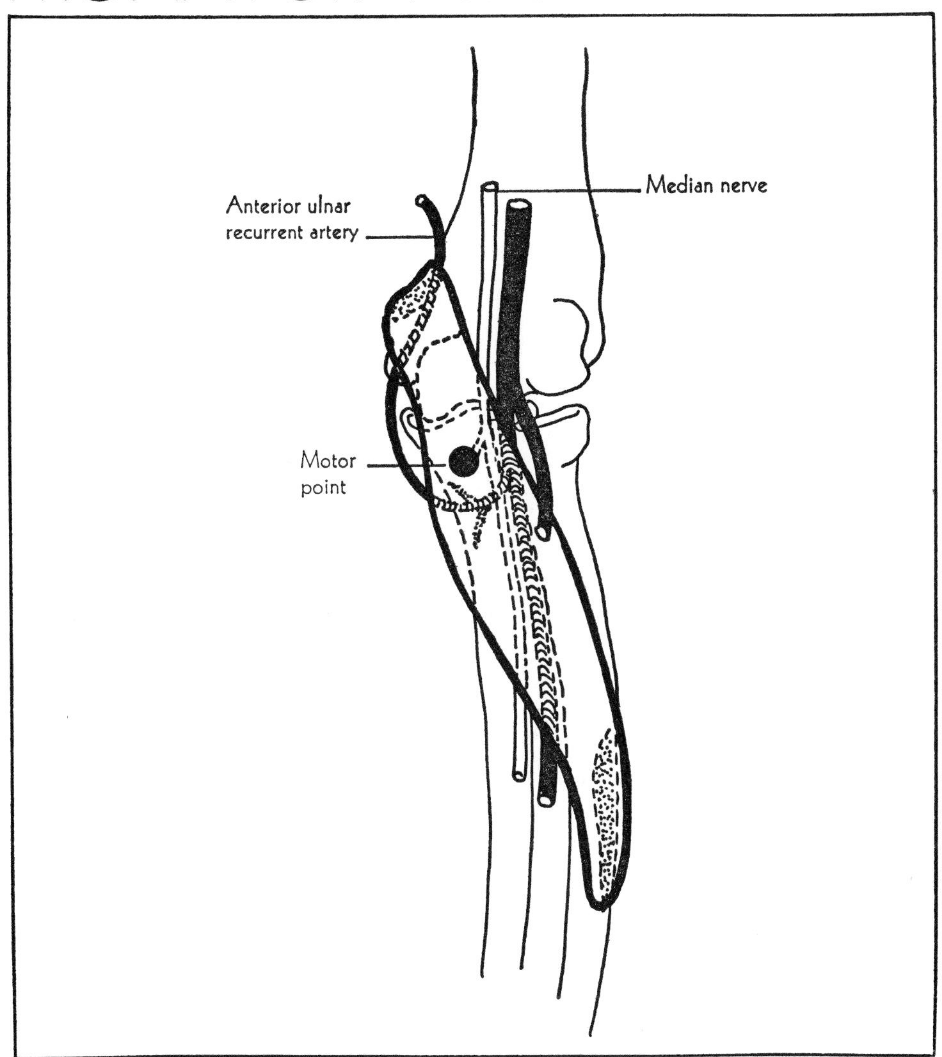

Median nerve
Anterior ulnar
recurrent artery
Motor
point

ORIGIN: Humeral head from medial epicondylar ridge of humerus and common flexor tendon, ulnar head from medial side of coronoid process of ulna

INSERTION: Middle of lateral surface of radius

FUNCTION: Pronates forearm; assists in flexing forearm

NERVE: Median

ARTERY: Anterior ulnar recurrent

REFERENCES:	GRAY	GRANT'S ATLAS
Muscle	463	59
Nerve	463, 961, 963, 971	60
Artery	622	Not shown

FLEXOR CARPI RADIALIS

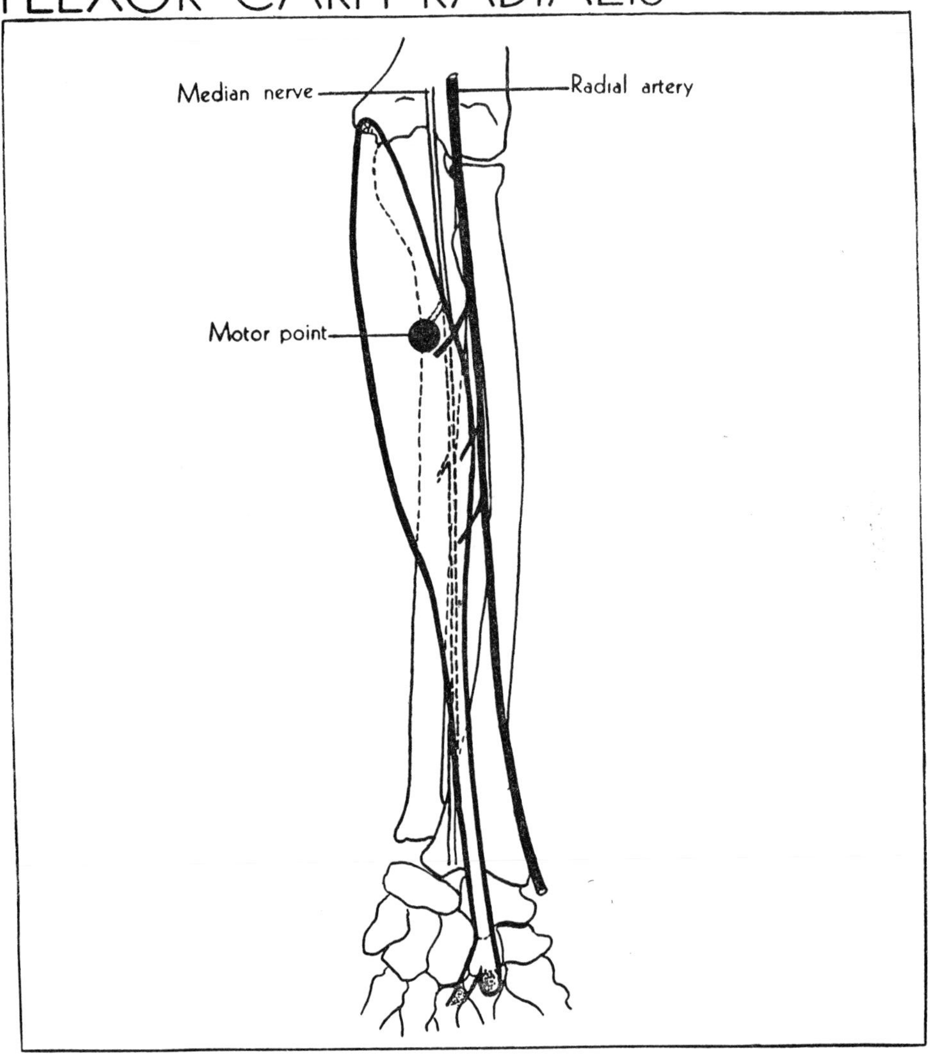

Median nerve
Radial artery
Motor point

ORIGIN:	Common flexor tendon from medial epicondyle of humerus, fascia of forearm
INSERTION:	Base of 2d and 3d metacarpal bones
FUNCTION:	Flexes wrist, assists in pronating and abducting hand, assists in flexing forearm
NERVE:	Median
ARTERY:	Muscular branches of radial

REFERENCES: GRAY		GRANT'S ATLAS
Muscle	463	59
Nerve	463, 961, 963, 971	Not shown
Artery	620	Not shown

PALMARIS LONGUS

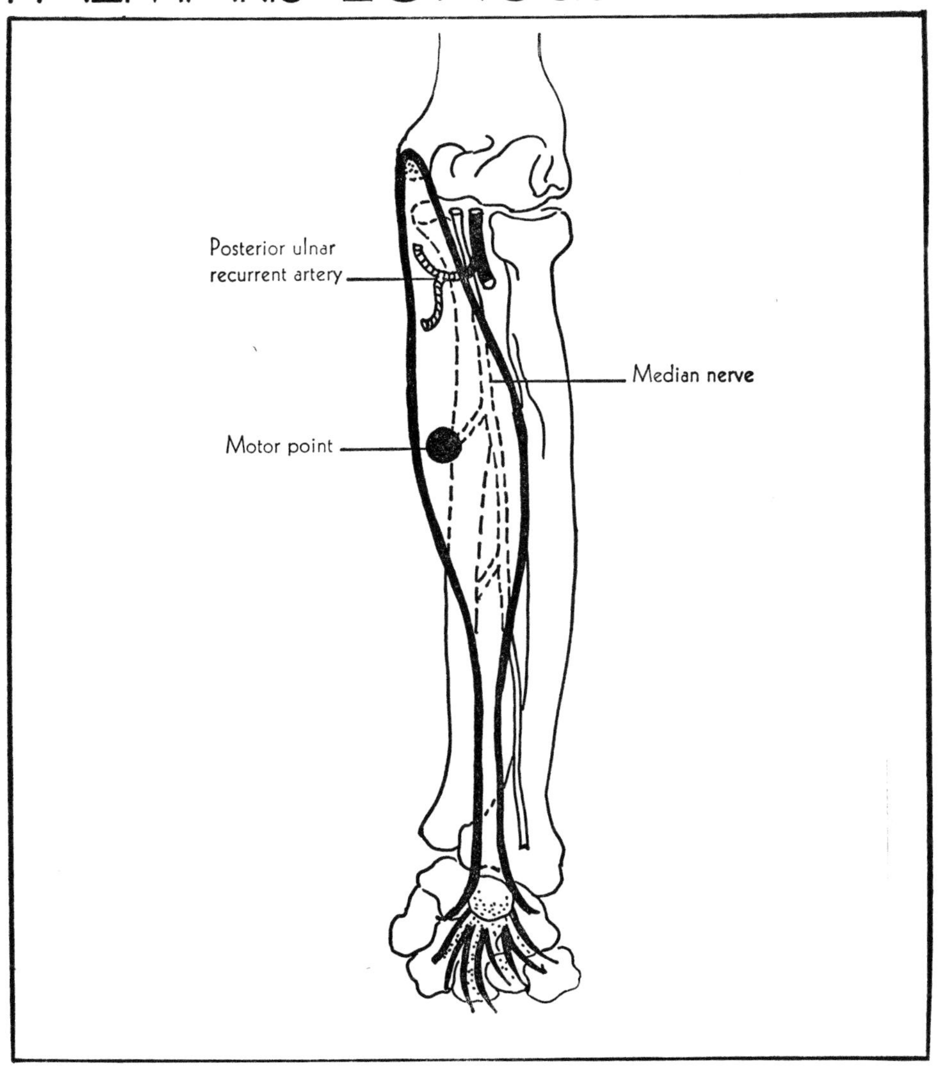

Posterior ulnar recurrent artery
Median nerve
Motor point

ORIGIN:	Common flexor tendon from medial epicondyle of humerus
INSERTION:	Transverse carpal ligament and palmar aponeurosis
FUNCTION:	Flexes wrist, assists in pronation and flexion of forearm
NERVE:	Median
ARTERY:	Posterior ulnar recurrent

REFERENCES:	GRAY	GRANT'S ATLAS
Muscle	463	59, 66
Nerve	463, 961, 963, 971	Not shown
Artery	622	Not shown

FLEXOR CARPI ULNARIS

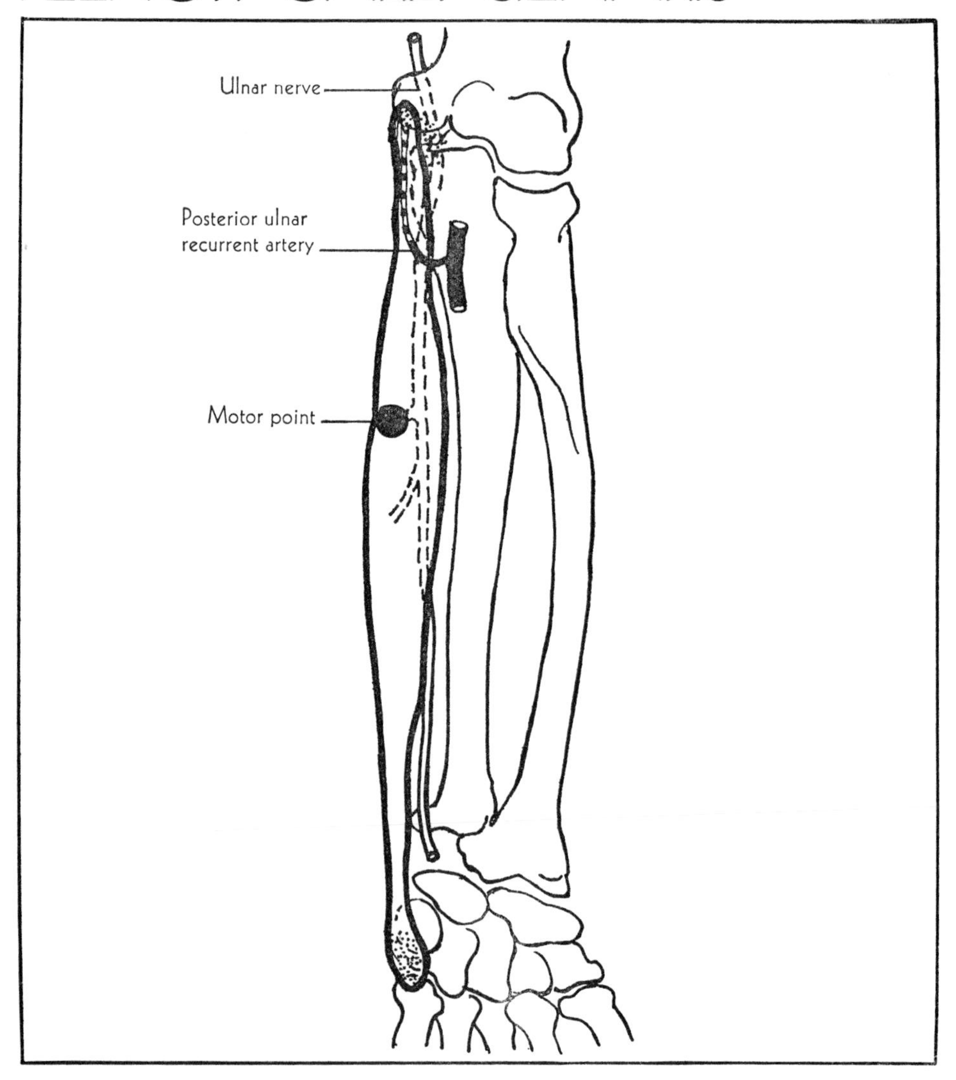

Ulnar nerve
Posterior ulnar recurrent artery
Motor point

ORIGIN: Humeral head from common flexor tendon from medial epicondyle of humerus, ulnar head from olecranon and dorsal border of ulna

INSERTION: Pisiform, hamate, 5th metacarpal bones

FUNCTION: Flexes wrist and assists in adducting it; assists in flexing forearm

NERVE: Ulnar

ARTERY: Posterior ulnar recurrent

REFERENCES:	GRAY	GRANT'S ATLAS
Muscle	463	59, 60
Nerve	464, 961, 963, 972	60, 61
Artery	622	**57, 61**

FLEXOR DIGITORUM SUPERFICIALIS

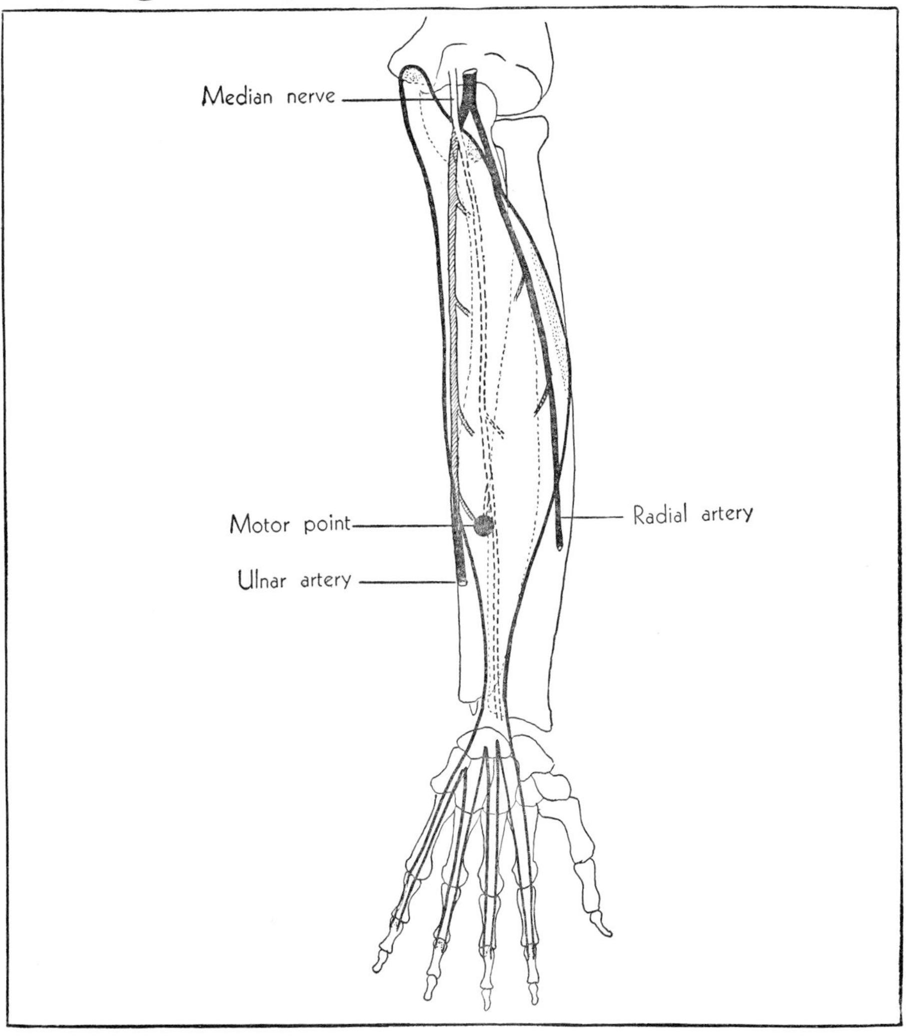

Median nerve
Motor point
Ulnar artery
Radial artery

ORIGIN: Humeral head from common flexor tendon from medial epicondyle of humerus, ulnar head from coronoid process of ulna, radial head from oblique line of radius

INSERTION: Margins of palmar surface of middle phalanx of medial 4 digits

FUNCTION: Flexes middle and proximal phalanges of medial 4 digits, aids in flexing wrist and forearm

NERVE: Median

ARTERY: Muscular branches of ulnar, muscular branches of radial

REFERENCES:	GRAY	GRANT'S ATLAS
Muscle	464	59, 66, 72
Nerve	464, 961, 963, 971	Not shown
Artery	620, 624	60

FLEXOR DIGITORUM PROFUNDUS

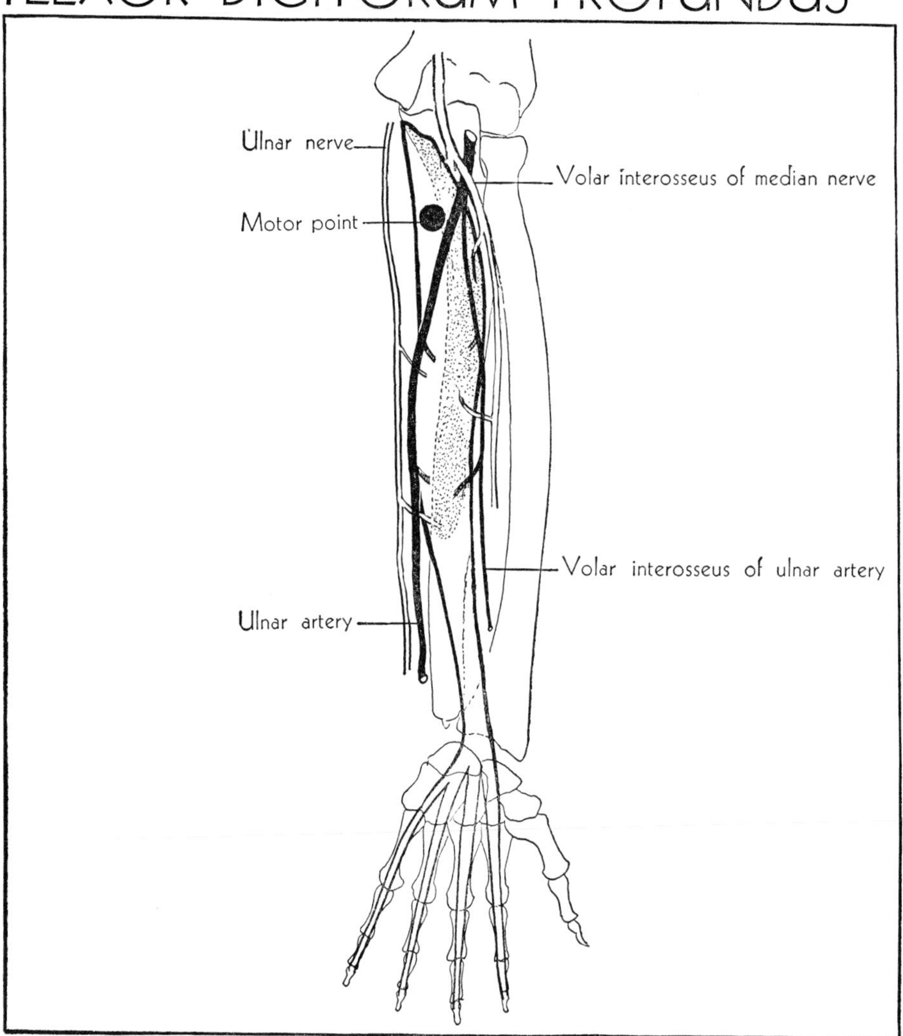

Ulnar nerve
Motor point
Volar interosseus of median nerve
Volar interosseus of ulnar artery
Ulnar artery

ORIGIN: Medial and anterior surface of ulna, interosseus membrane, deep fascia of forearm

INSERTION: Distal phalanges of medial 4 digits

FUNCTION: Flexes terminal phalanges of medial 4 digits after superficialis flexes 2d phalanges, aids in flexing wrist

NERVE: Ulnar, volar interosseus of median

ARTERY: Volar interosseus of ulnar, muscular branches of ulnar

REFERENCES:	GRAY	GRANT'S ATLAS
Muscle	464	61, 72
Nerve	465, 961, 963, 971, 972	60, 61
Artery	**622, 624**	**61**

FLEXOR POLLICIS LONGUS

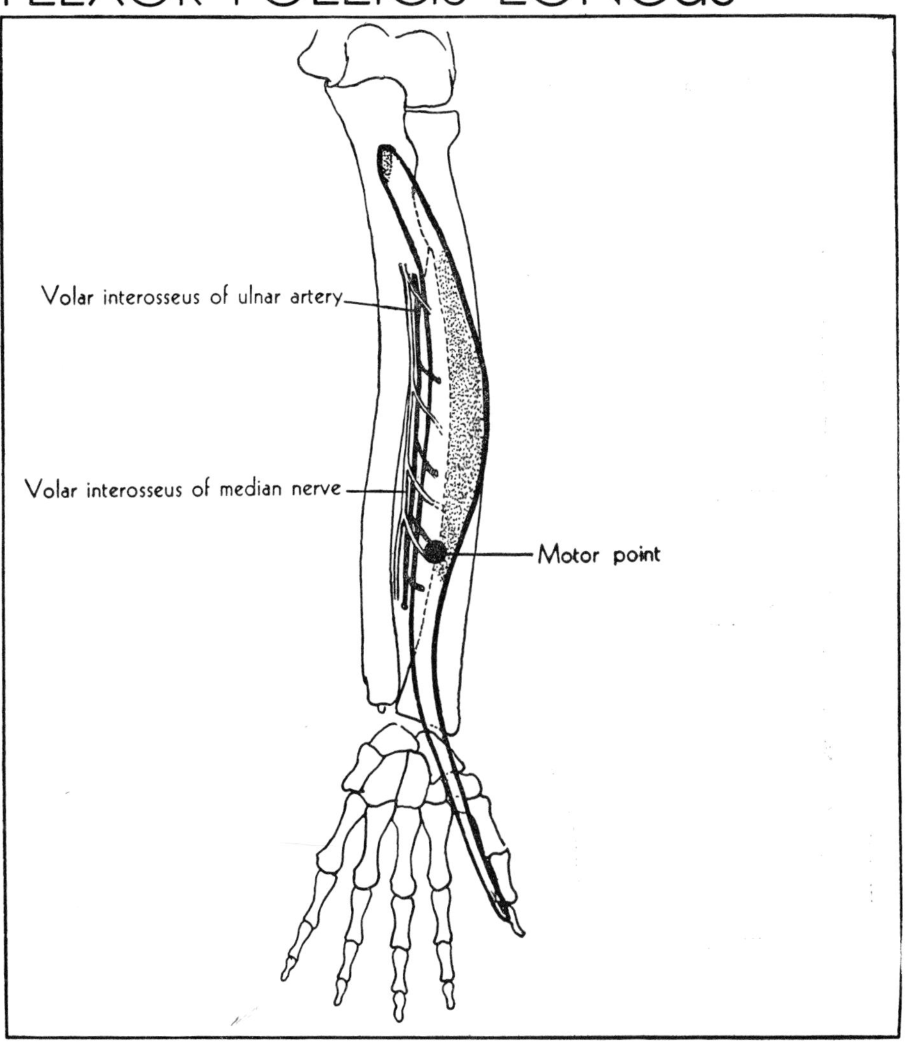
Volar interosseus of ulnar artery
Volar interosseus of median nerve
Motor point

ORIGIN:	Volar surface of radius, adjacent interosseus membrane, medial border of coronoid process of ulna
INSERTION:	Base of distal phalanx of thumb on palmar surface
FUNCTION:	Flexes thumb
NERVE:	Volar interosseus of median
ARTERY:	Volar interosseus of ulnar

REFERENCES: GRAY		GRANT'S ATLAS
Muscle	466	61, 71
Nerve	466, 961, 963, 971	61
Artery	622	61

PRONATOR QUADRATUS

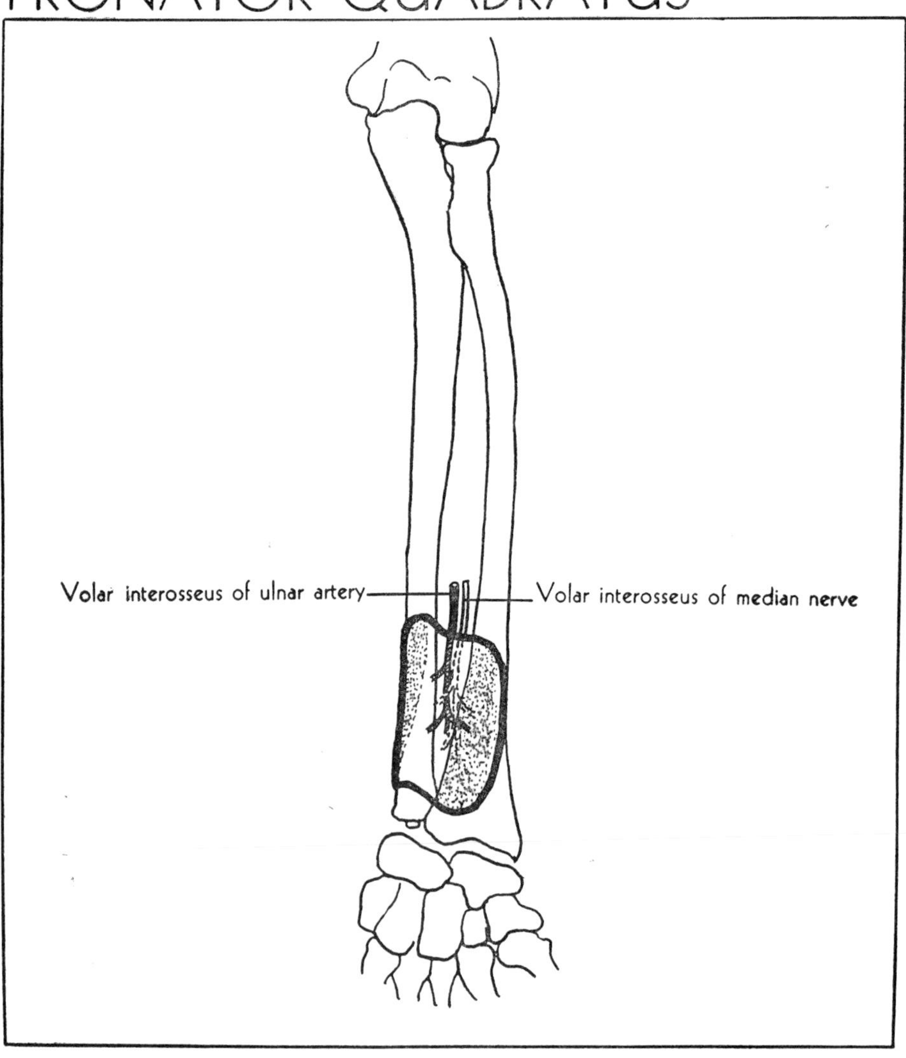
Volar interosseus of ulnar artery
Volar interosseus of median nerve

ORIGIN: Distal fourth of volar surface of ulna
INSERTION: Distal fourth of lateral border on volar surface of radius
FUNCTION: Pronates forearm
NERVE: Volar interosseus of median
ARTERY: Volar interosseus of ulnar

REFERENCES:	GRAY	GRANT'S ATLAS
Muscle	467	72
Nerve	467, 961, 963, 971	Not shown
Artery	622	Not shown

BRACHIORADIALIS

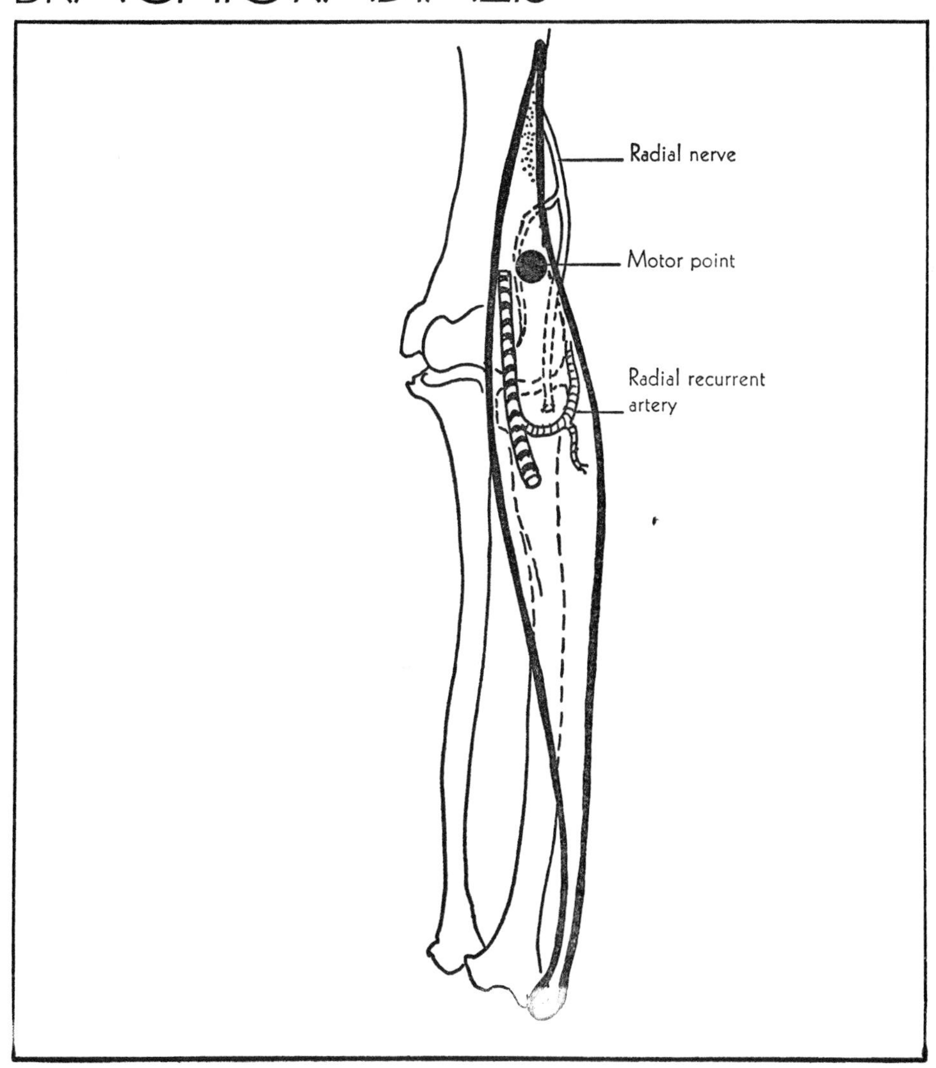
Radial nerve
Motor point
Radial recurrent
artery

ORIGIN: Proximal two-thirds of lateral supracondylar ridge of humerus, lateral intermuscular septum

INSERTION: Lateral side of base of styloid process of radius

FUNCTION: Flexes forearm after flexion has been started by biceps and brachialis; may also act as a semipronator and semisupinator

NERVE: Radial

ARTERY: Radial recurrent

REFERENCES:	GRAY	GRANT'S ATLAS
Muscle	467	59
Nerve	468, 961, 963, 978	61
Artery	620	60, 61

EXTENSOR CARPI RADIALIS LONGUS

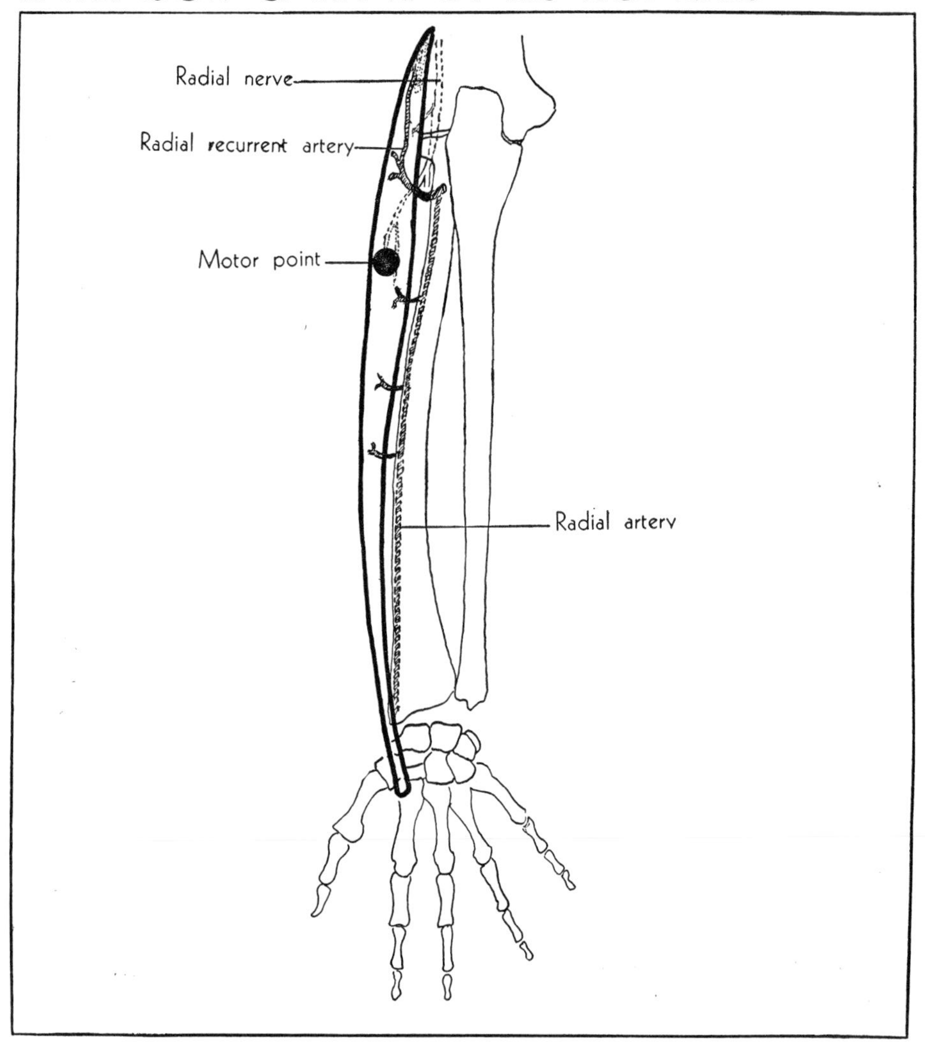
Radial nerve
Radial recurrent artery
Motor point
Radial artery

ORIGIN: Lower third of lateral supracondylar ridge of humerus, lateral intermuscular septum

INSERTION: Dorsal surface of base of 2d metacarpal bone

FUNCTION: Extends wrist, abducts hand

NERVE: Radial

ARTERY: Muscular branches of radial, radial recurrent

REFERENCES:	GRAY	GRANT'S ATLAS
Muscle	468	78, 87
Nerve	468, 961, 963, 978	Not shown
Artery	620	61

EXTENSOR CARPI RADIALIS BREVIS

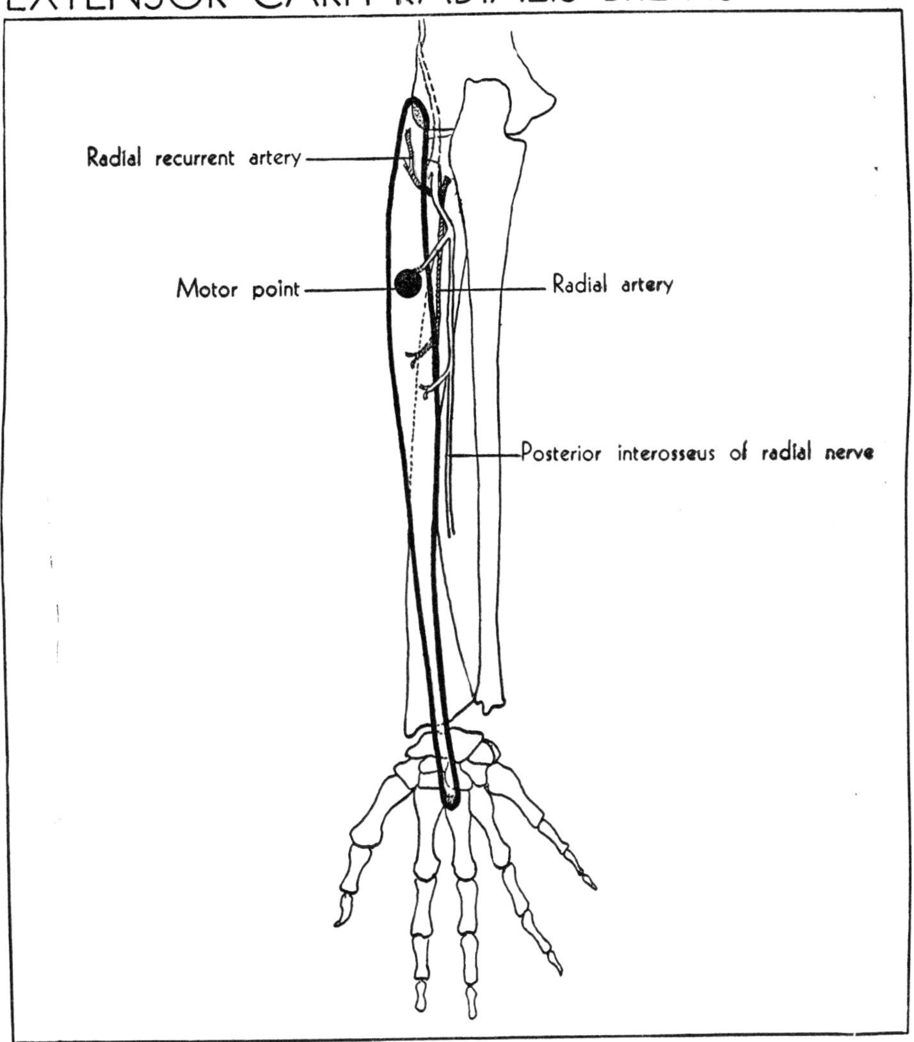
Radial recurrent artery
Motor point
Radial artery
Posterior interosseus of radial nerve

ORIGIN: From common extensor tendon from lateral epicondyle of humerus, radial collateral ligament of elbow joint, intermuscular septa

INSERTION: Dorsal surface of base of 3d metacarpal bone

FUNCTION: Extends wrist, abducts hand

NERVE: Posterior interosseus of radial

ARTERY: Muscular branches of radial, radial recurrent

REFERENCES: GRAY	GRANT'S ATLAS
Muscle 468	78, 87
Nerve 469, 961, 963, 978	61
Artery 620	61

EXTENSOR DIGITORUM

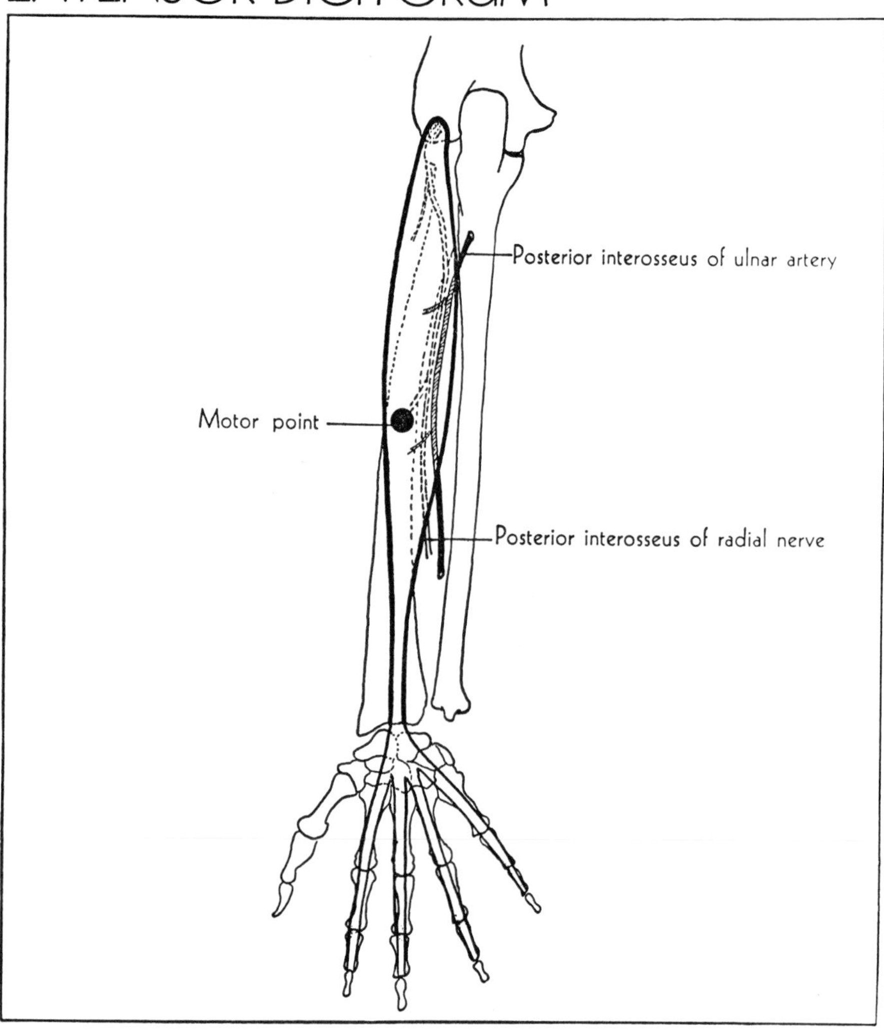

Posterior interosseus of ulnar artery
Motor point
Posterior interosseus of radial nerve

ORIGIN: Lateral epicondyle of humerus by common extensor tendon, intermuscular septa

INSERTION: Lateral and dorsal surface of phalanges of medial 4 digits

FUNCTION: Extends medial 4 digits; assists in extension of wrist

NERVE: Posterior interosseus of radial

ARTERY: Posterior interosseus of ulnar

REFERENCES: GRAY		GRANT'S ATLAS
Muscle	469	78, 87
Nerve	469, 961, 963, 978	79
Artery	622	79

EXTENSOR DIGITI MINIMI

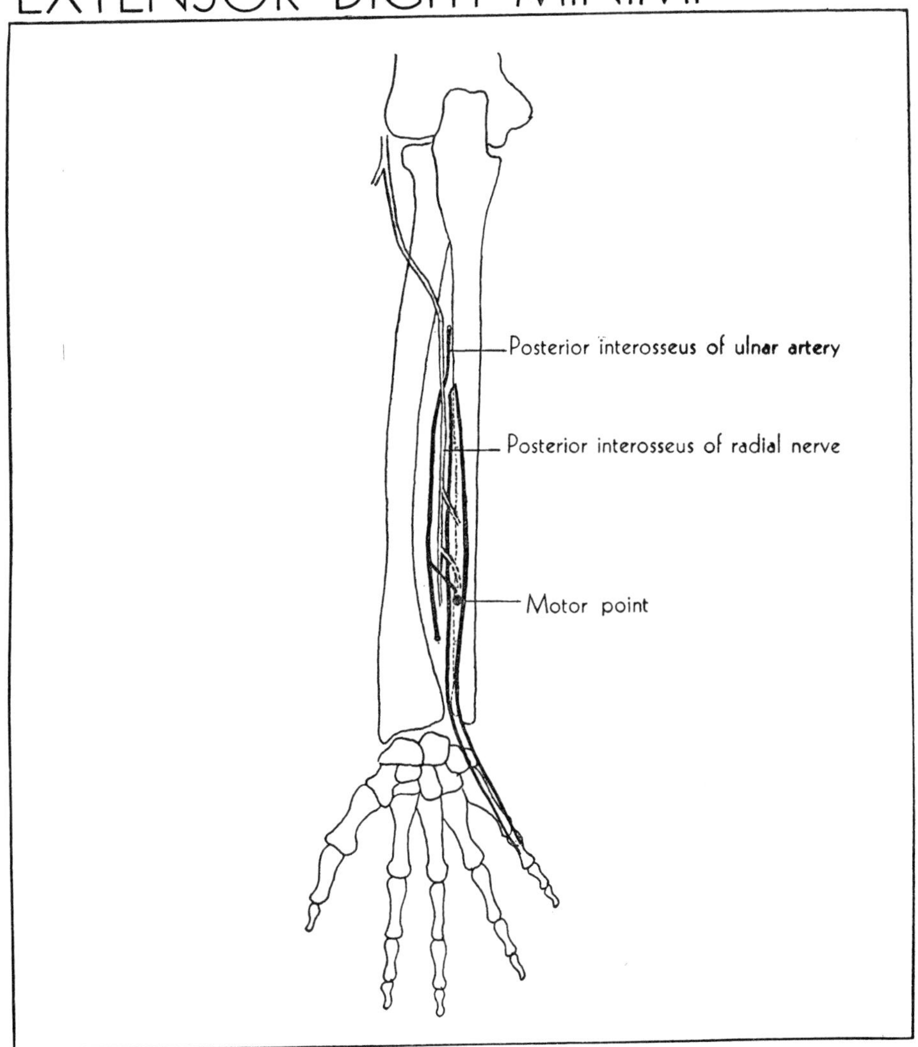
Posterior interosseus of ulnar artery
Posterior interosseus of radial nerve
Motor point

ORIGIN: Common extensor tendon from lateral epicondyle of the humerus, intermuscular septa

INSERTION: Dorsum of proximal phalanx of 5th digit

FUNCTION: Extends 5th digit

NERVE: Posterior interosseus of radial

ARTERY: Posterior interosseus of ulnar

REFERENCES: GRAY	GRANT'S ATLAS
Muscle 469	78, 87
Nerve 470, 961, 963, 978	79
Artery 622	79

EXTENSOR CARPI ULNARIS

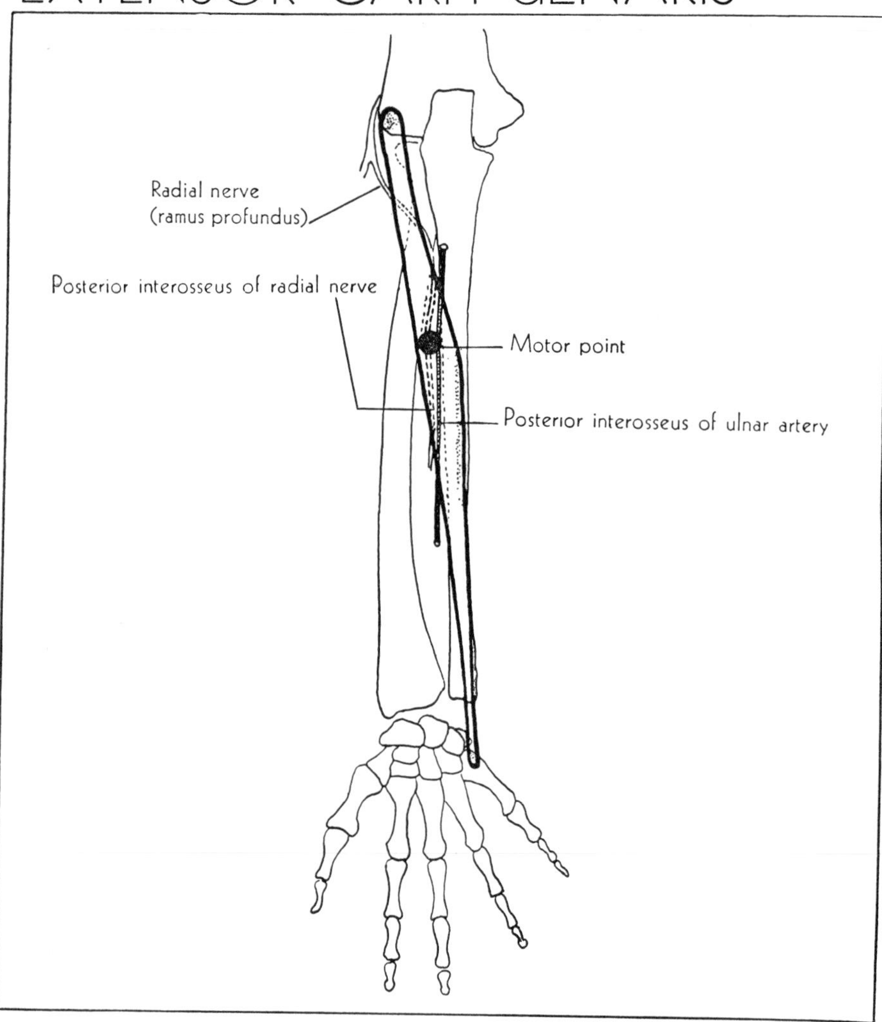

Radial nerve
(ramus profundus)
Posterior interosseus of radial nerve
Motor point
Posterior interosseus of ulnar artery

ORIGIN: From common extensor tendon from lateral epicondyle of humerus, and from posterior border of ulna

INSERTION: Medial side of base of 5th metacarpal bone

FUNCTION: Extends wrist, adducts hand

NERVE: Posterior interosseus of radial

ARTERY: Posterior interosseus of ulnar

REFERENCES:	GRAY	GRANT'S ATLAS
Muscle	470	78
Nerve	470, 961, 963, 978	79
Artery	622	79

ANCONEUS

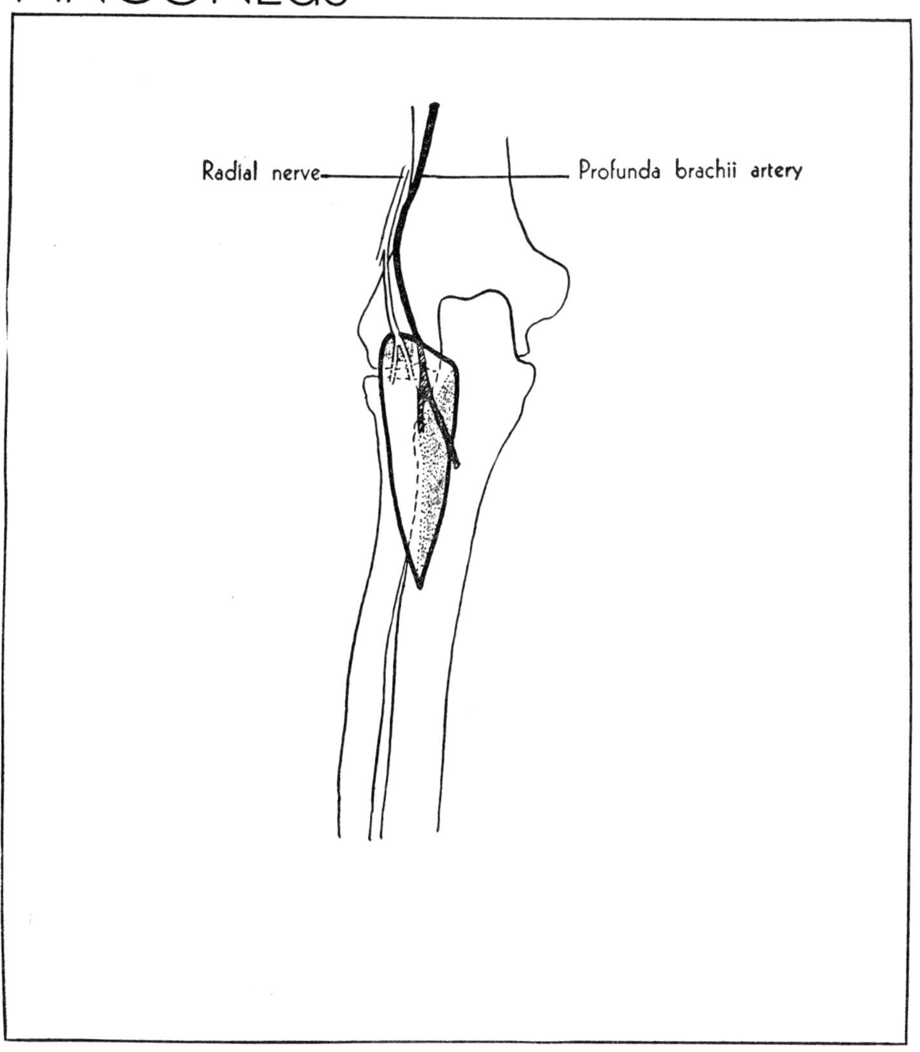
Radial nerve
Profunda brachii artery

ORIGIN: Lateral epicondyle of humerus, posterior ligament of elbow joint
INSERTION: Lateral side of olecranon and posterior surface of ulna
FUNCTION: Assists triceps in extending forearm
NERVE: Radial
ARTERY: Branch of profunda brachii

REFERENCES:	GRAY	GRANT'S ATLAS
Muscle	470	56
Nerve	470, 961, 963, 977	78
Artery	616	Not shown

SUPINATOR

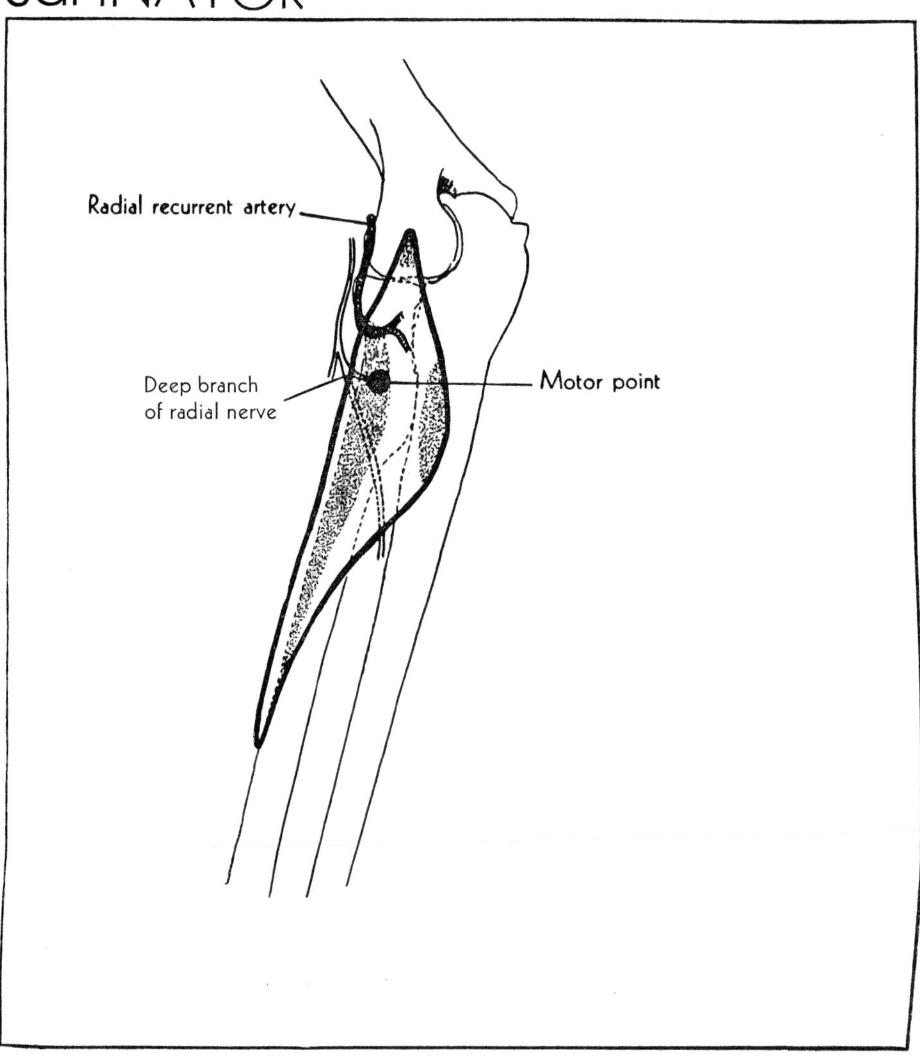
Radial recurrent artery
Deep branch
of radial nerve
Motor point

ORIGIN: Lateral epicondyle of humerus, lateral ligament of elbow joint and annular ligament of radius, supinator crest and fossa of ulna

INSERTION: Lateral and anterior surface of radius in its upper third

FUNCTION: Supinates forearm

NERVE: Deep branch of radial

ARTERY: Radial recurrent; post-interosseous of ulnar

REFERENCES: GRAY		GRANT'S ATLAS
Muscle	470	46, 79
Nerve	471, 961, 963, 978	61
Artery	620, 622	61

ABDUCTOR POLLICIS LONGUS

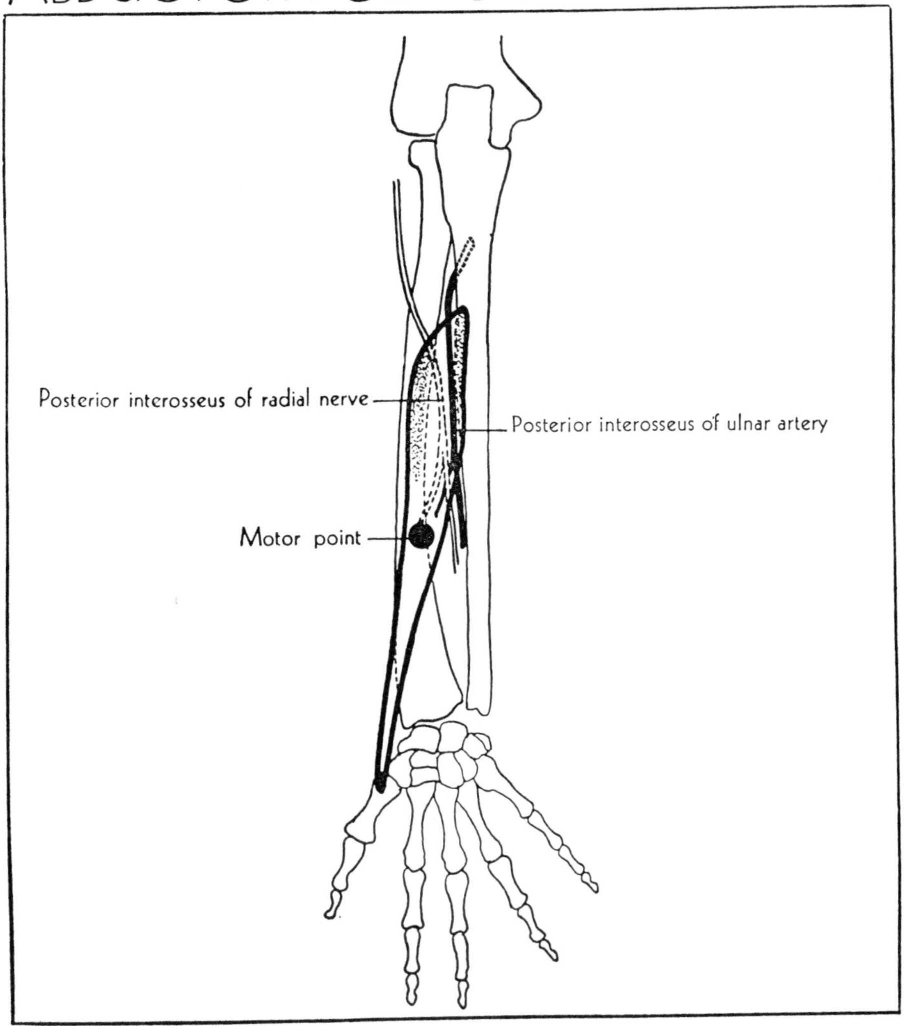
Posterior interosseus of radial nerve
Posterior interosseus of ulnar artery
Motor point

ORIGIN: Posterior surface of ulna, interosseus membrane, middle third of posterior surface of radius

INSERTION: Radial side of base of 1st metacarpal bone

FUNCTION: Abducts thumb and wrist

NERVE: Posterior interosseus of radial

ARTERY: Posterior interosseus of ulnar

REFERENCES: GRAY		GRANT'S ATLAS
Muscle	471	67, 78, 79
Nerve	471, 961, 963, 978	79
Artery	622	79

EXTENSOR POLLICIS BREVIS

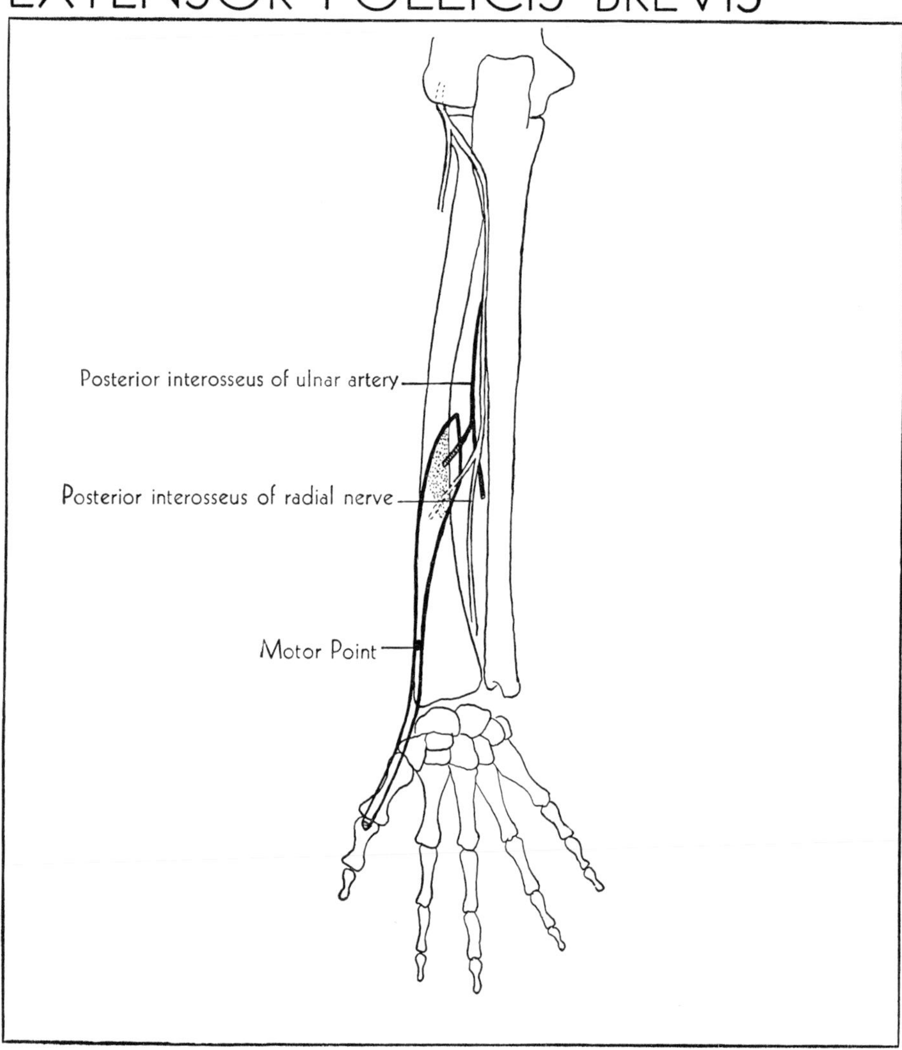

Posterior interosseus of ulnar artery
Posterior interosseus of radial nerve
Motor Point

ORIGIN:	Posterior surface of radius, interosseus membrane
INSERTION:	Base of proximal phalanx of thumb
FUNCTION:	Extends proximal phalanx of thumb
NERVE:	Posterior interosseus of radial
ARTERY:	Posterior interosseus of ulnar

REFERENCES:	GRAY	GRANT'S ATLAS
Muscle	471	78, 79
Nerve	472, 961, 963, 978	79
Artery	622	Not shown

EXTENSOR POLLICIS LONGUS

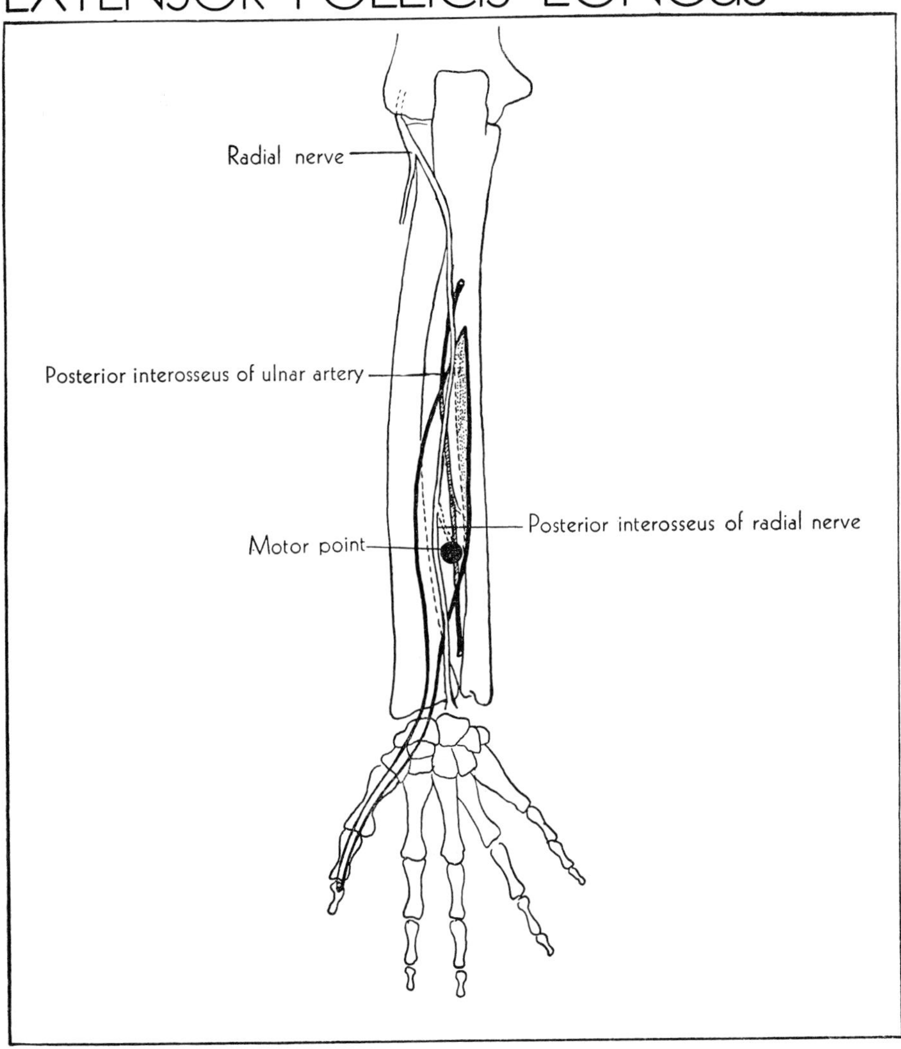

Radial nerve
Posterior interosseus of ulnar artery
Posterior interosseus of radial nerve
Motor point

ORIGIN:	Middle third of posterior surface of ulna, interosseus membrane
INSERTION:	Base of distal phalanx of thumb
FUNCTION:	Extends terminal phalanx of thumb
NERVE:	Posterior interosseus of radial
ARTERY:	Posterior interosseus of ulnar

REFERENCES: GRAY	GRANT'S ATLAS
Muscle 472	78, 79
Nerve 472, 961, 963, 978	79
Artery 622	Not shown

EXTENSOR INDICIS

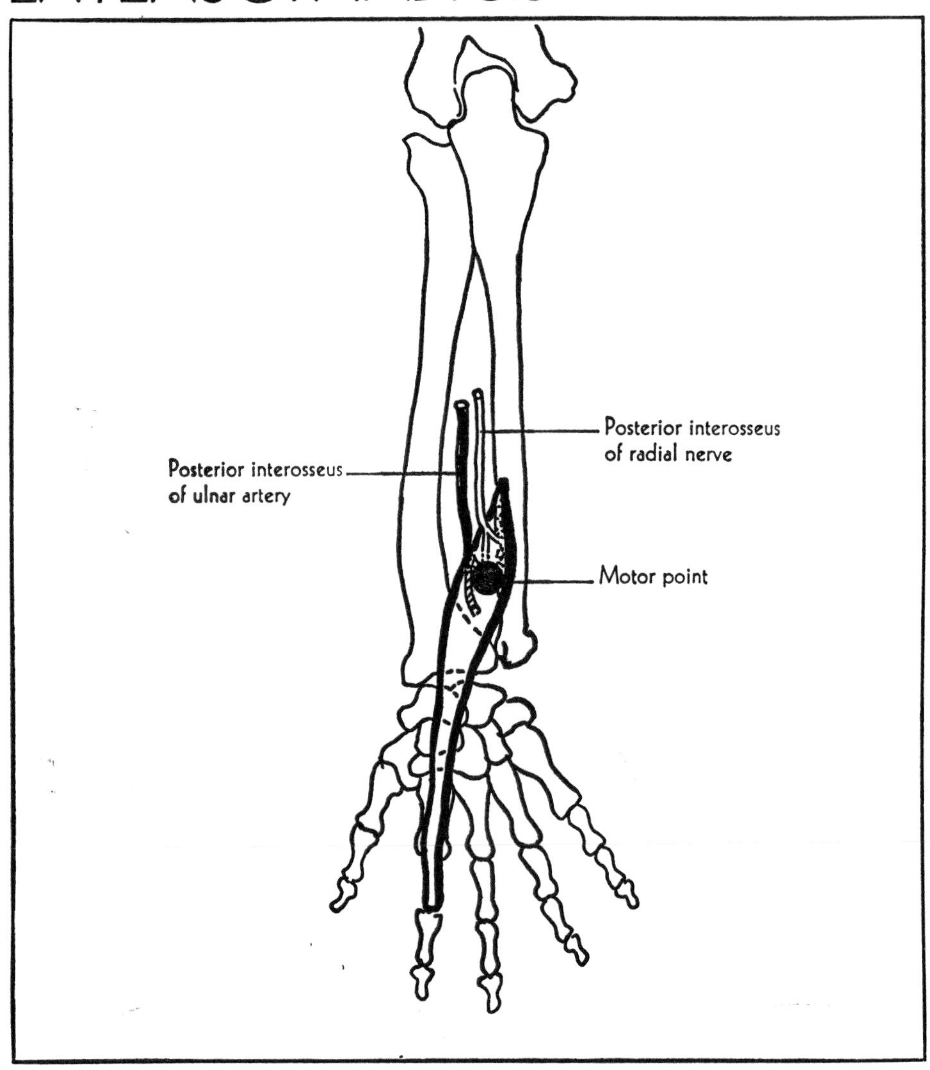
Posterior interosseus of radial nerve
Posterior interosseus of ulnar artery
Motor point

ORIGIN: Posterior surface of ulna, interosseus membrane
INSERTION: Dorsum of proximal phalanx of index finger
FUNCTION: Extends proximal phalanx of index finger
NERVE: Posterior interosseus of radial
ARTERY: Posterior interosseus of ulnar

REFERENCES:	GRAY	GRANT'S ATLAS
Muscle	472	79, 87
Nerve	472, 961, 973, 978	79
Artery	622	Not shown

ABDUCTOR POLLICIS BREVIS

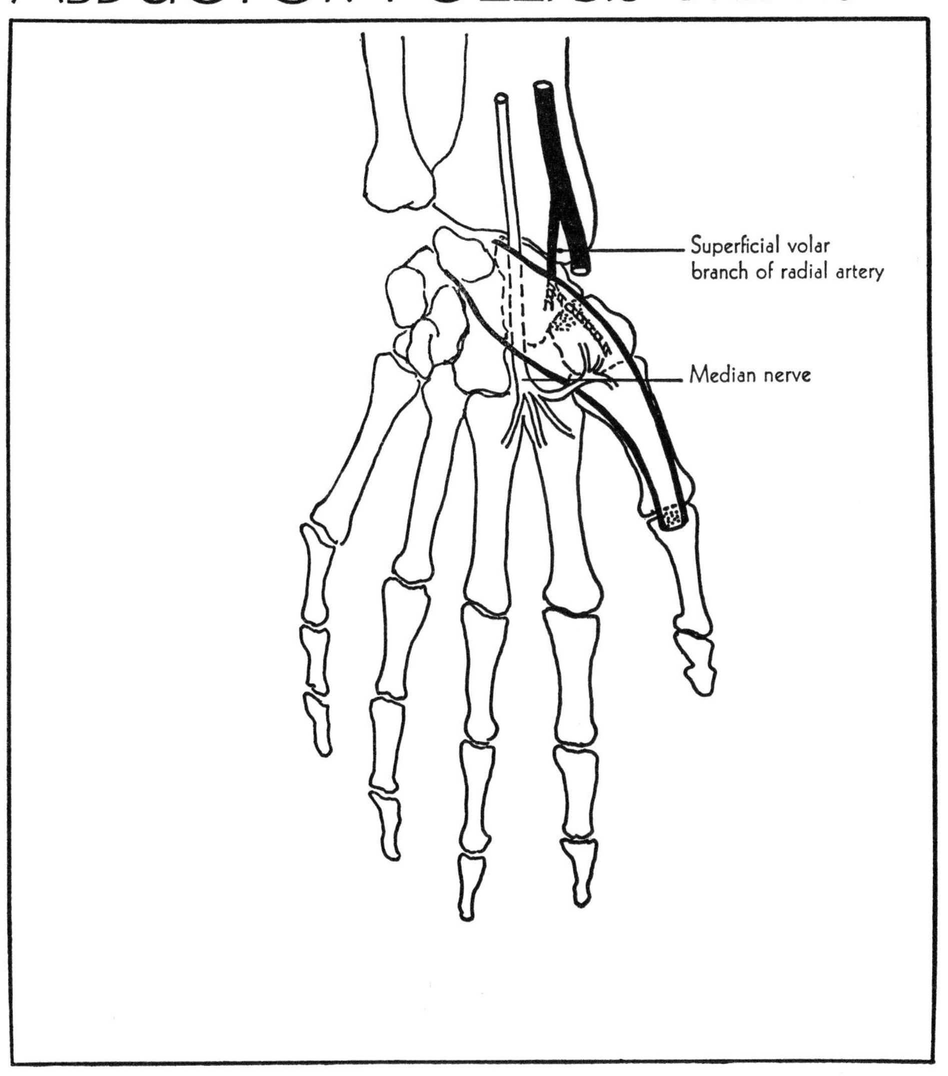
Superficial volar
branch of radial artery
Median nerve

ORIGIN: Transverse carpal ligament, scaphoid and trapezium bones
INSERTION: Radial side of base of proximal phalanx of thumb
FUNCTION: Abducts thumb, draws thumb forward at right angles to palm
NERVE: Muscular branches of median
ARTERY: Superficial volar branch of radial

REFERENCES: GRAY	GRANT'S ATLAS
Muscle 482	67
Nerve 482, 961, 963, 971	67
Artery 20	67

OPPONENS POLLICIS

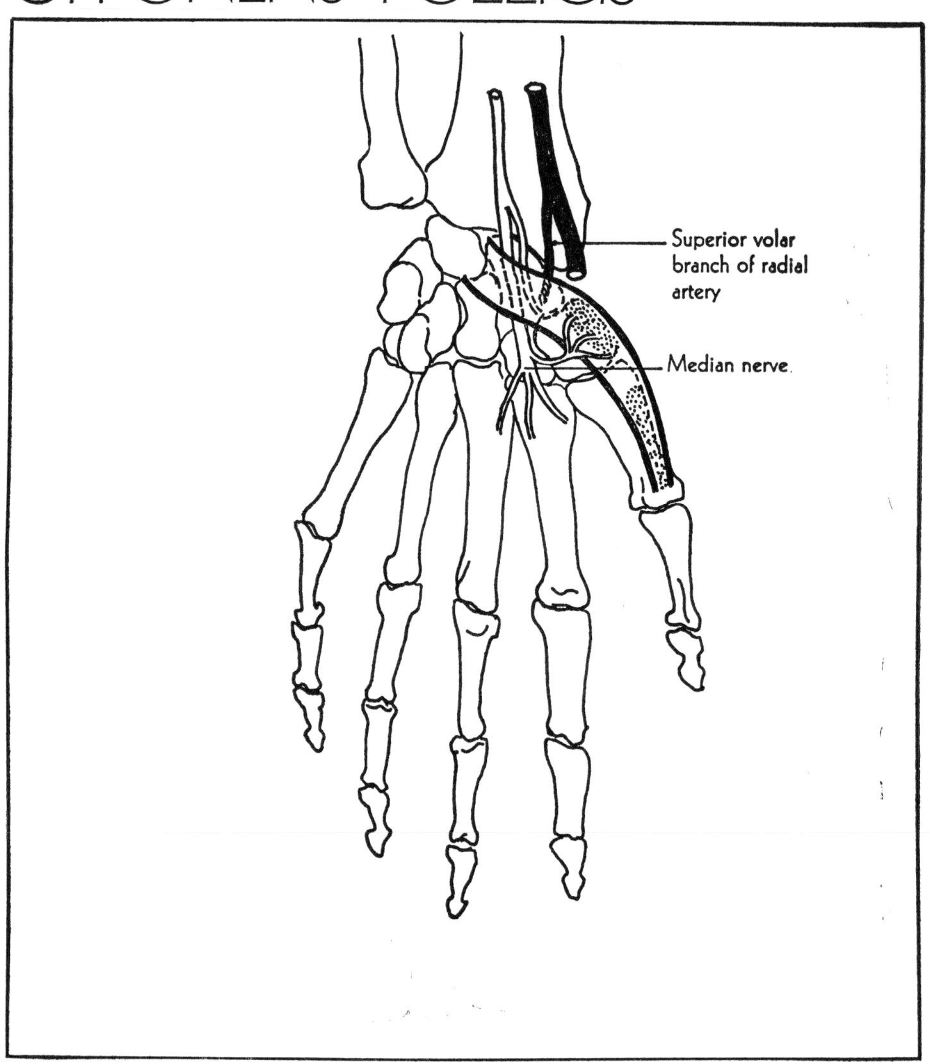
Superior volar
branch of radial
artery
Median nerve

ORIGIN: Transverse carpal ligament, trapezium bone

INSERTION: Anterior surface, radial side of 1st metacarpal bone

FUNCTION: Draws 1st metacarpal bone forwards, and medially, opposing thumb to each of the other digits

NERVE: Muscular branches of median

ARTERY: Superficial volar branch of radial

REFERENCES:	GRAY	GRANT'S ATLAS
Muscle	482	61, 68
Nerve	482, 961, 963, 971	68
Artery	620	62.1

FLEXOR POLLICIS BREVIS

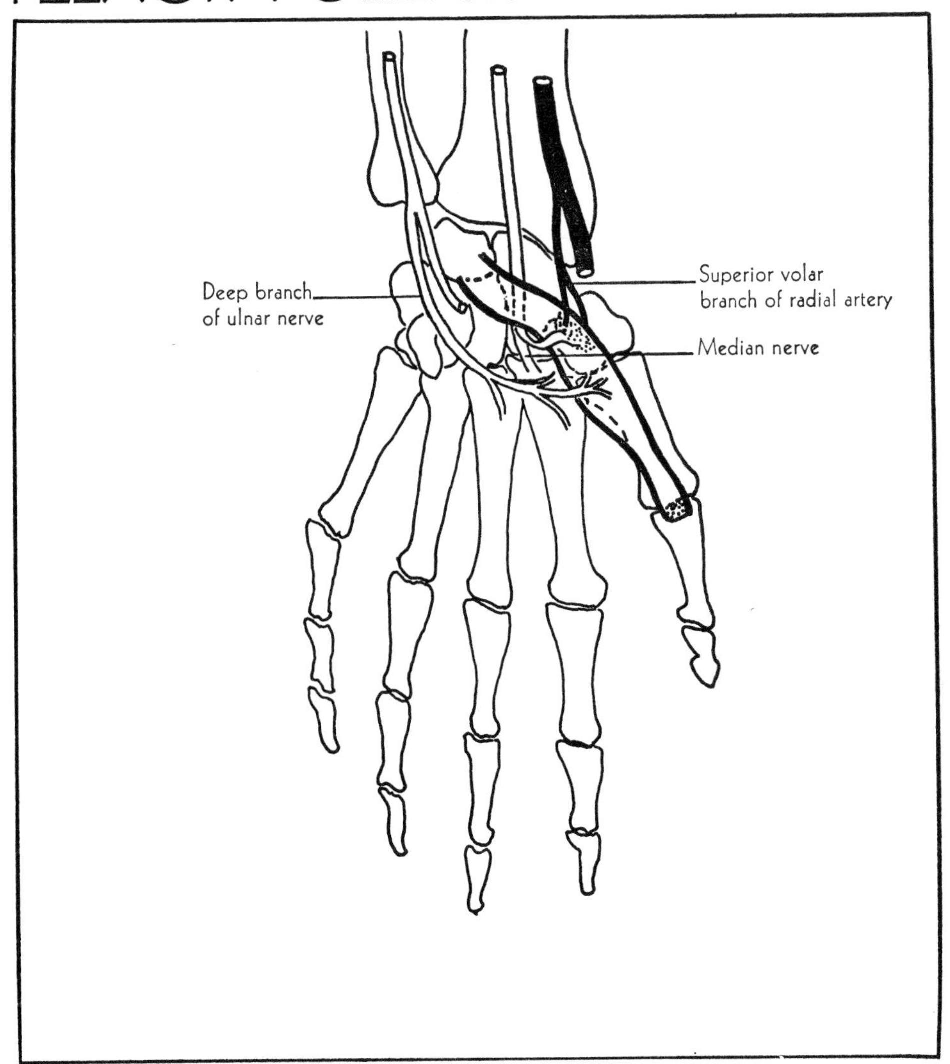

Deep branch
of ulnar nerve
Superior volar
branch of radial artery
Median nerve

ORIGIN: Transverse carpal ligament, trapezium bone
INSERTION: Base of proximal phalanx of thumb
FUNCTION: Flexes proximal phalanx of thumb
NERVE: Muscular branches of median; deep branch of ulnar
ARTERY: Superficial volar branch of radial

REFERENCES: GRAY		GRANT'S ATLAS
Muscle	482	61, 68
Nerve	483, 961, 963, 971, 975	68
Artery	620	67

ADDUCTOR POLLICIS

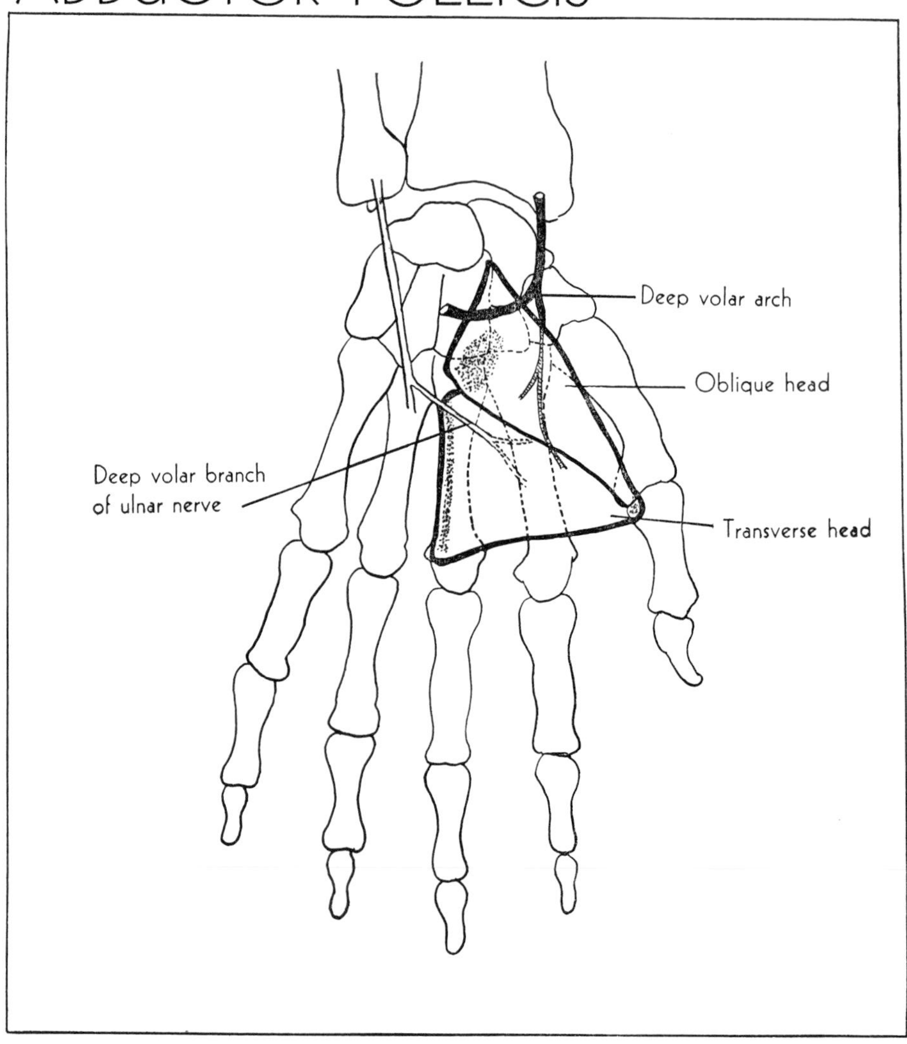
Deep volar arch
Oblique head
Deep volar branch
of ulnar nerve
Transverse head

ORIGIN: Oblique head from trapezium, trapezoid and capitate bones, base of 2d and 3d metacarpal bone, transverse head from volar surface of 3d metacarpal bone

INSERTION: Ulnar side of base of proximal phalanx of thumb

FUNCTION: Adducts thumb, aids in opposition

NERVE: Deep volar branch of ulnar

ARTERY: Deep volar arch

REFERENCES:	GRAY	GRANT'S ATLAS
Muscle	483	70, 71, 74
Nerve	483, 961, 963, 975	71
Artery	620	71

PALMARIS BREVIS

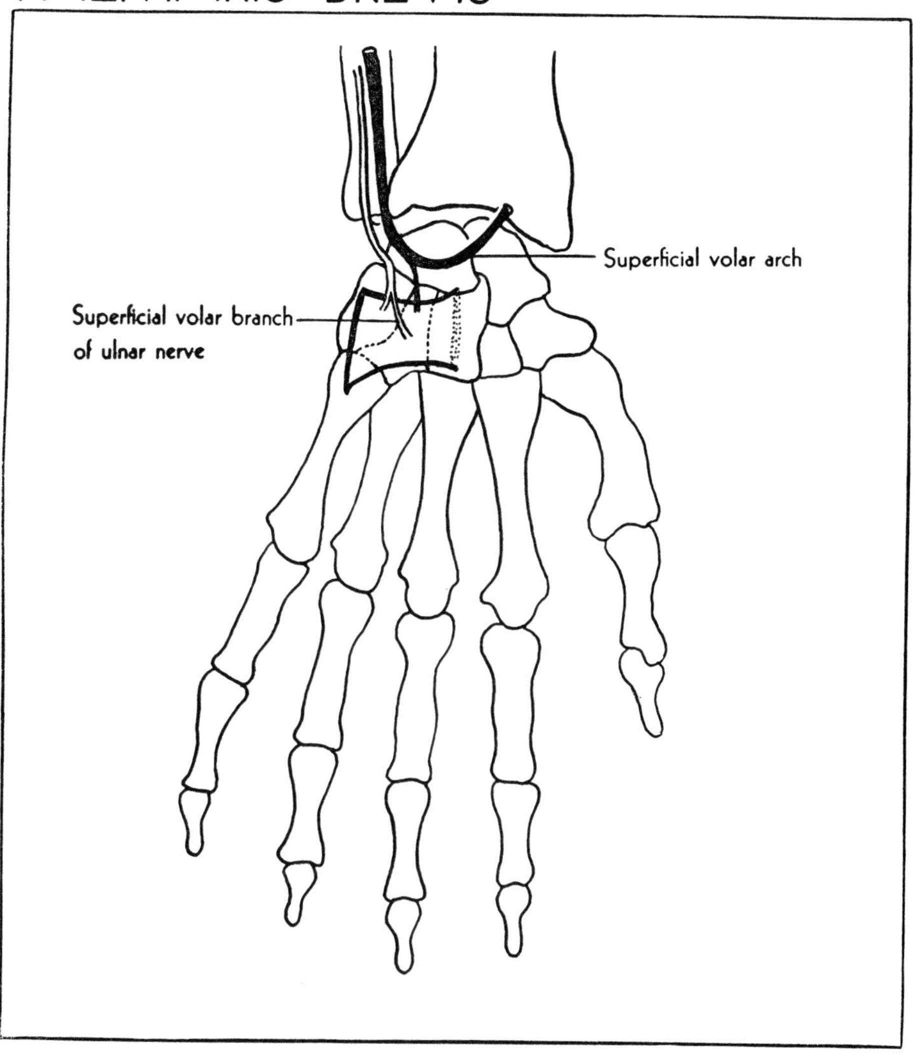

Superficial volar arch
Superficial volar branch
of ulnar nerve

ORIGIN: Ulnar side of transverse carpal ligament, palmar aponeurosis
INSERTION: Skin on ulnar border of palm
FUNCTION: Corrugates skin on ulnar side of palm, deepens the hollow of the hand
NERVE: Superficial volar branch of ulnar
ARTERY: Superficial volar arch

REFERENCES:	GRAY	GRANT'S ATLAS
Muscle	484	67
Nerve	485, 961, 963, 974	Not shown
Artery	620	Not shown

ABDUCTOR DIGITI MINIMI

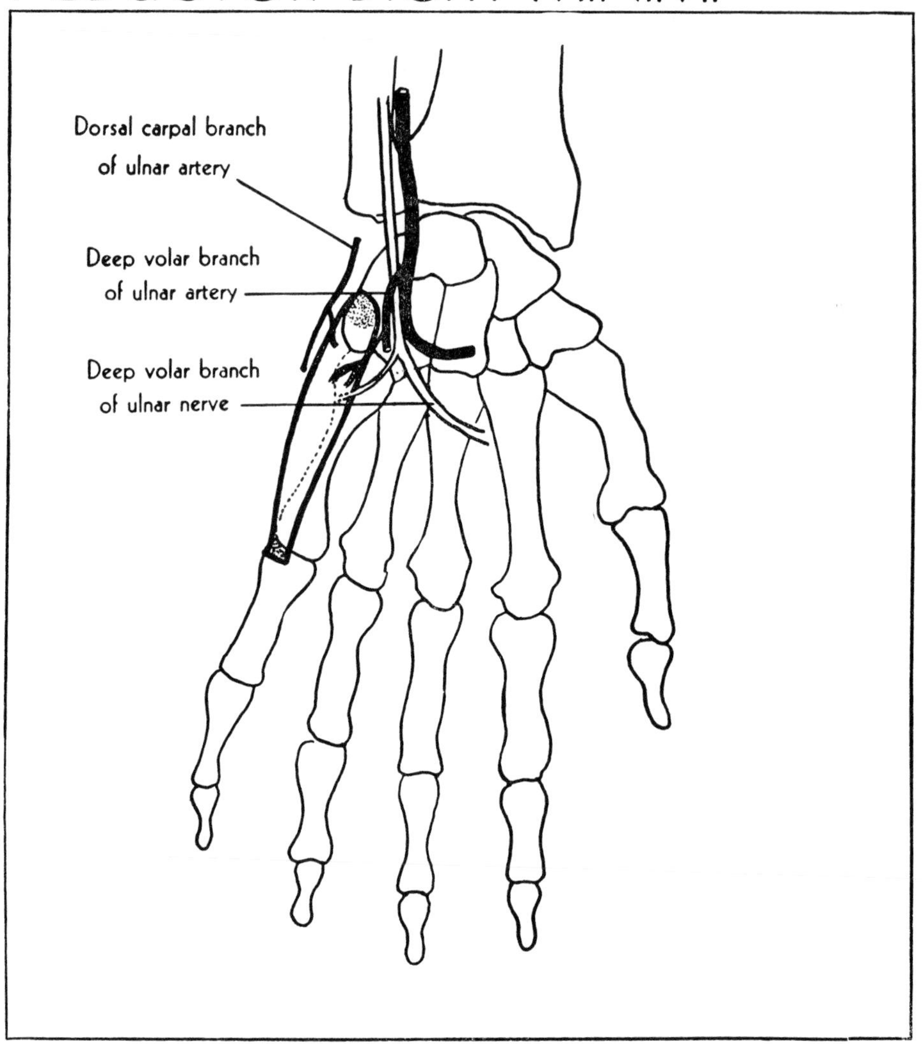

Dorsal carpal branch of ulnar artery
Deep volar branch of ulnar artery
Deep volar branch of ulnar nerve

ORIGIN:	Pisiform bone, tendon of flexor carpi ulnaris
INSERTION:	Medial side of base of proximal phalanx of 5th digit and aponeurosis of extensor digiti minimi
FUNCTION:	Abducts 5th digit from 4th digit
NERVE:	Deep volar branch of ulnar
ARTERY:	Deep volar branch of ulnar, dorsal carpal branch of ulnar

REFERENCES:	GRAY	GRANT'S ATLAS
Muscle	485	67, 81
Nerve	485, 961, 963, 975	68
Artery	624	71, 81

FLEXOR DIGITI MINIMI BREVIS

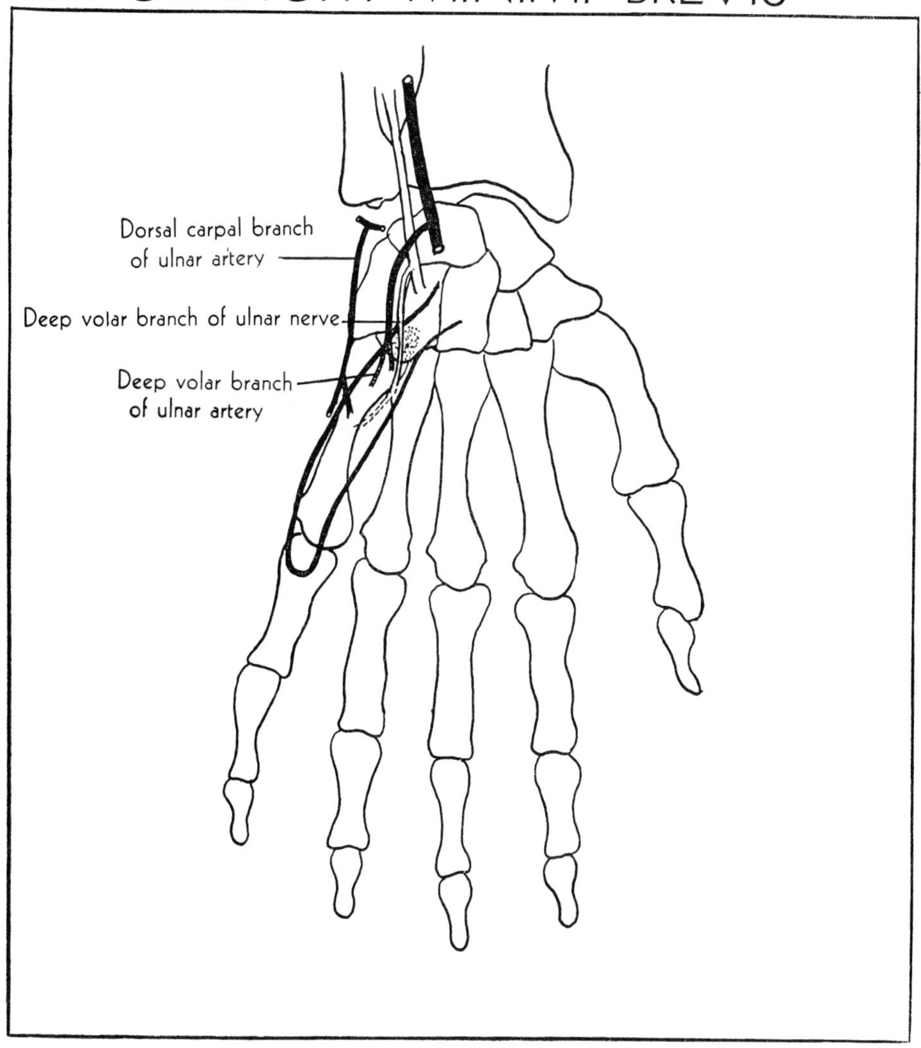
Dorsal carpal branch
of ulnar artery
Deep volar branch of ulnar nerve
Deep volar branch
of ulnar artery

ORIGIN: Transverse carpal ligament, hamulus of hamate bone
INSERTION: Ulnar side, base of proximal phalanx of 5th digit
FUNCTION: Flexes proximal phalanx of 5th digit
NERVE: Deep volar branch of ulnar
ARTERY: Deep volar branch of ulnar, dorsal carpal branch of ulnar

REFERENCES:	GRAY	GRANT'S ATLAS
Muscle	485	68
Nerve	485, 961, 963, 975	68
Artery	624	71

OPPONENS DIGITI MINIMI

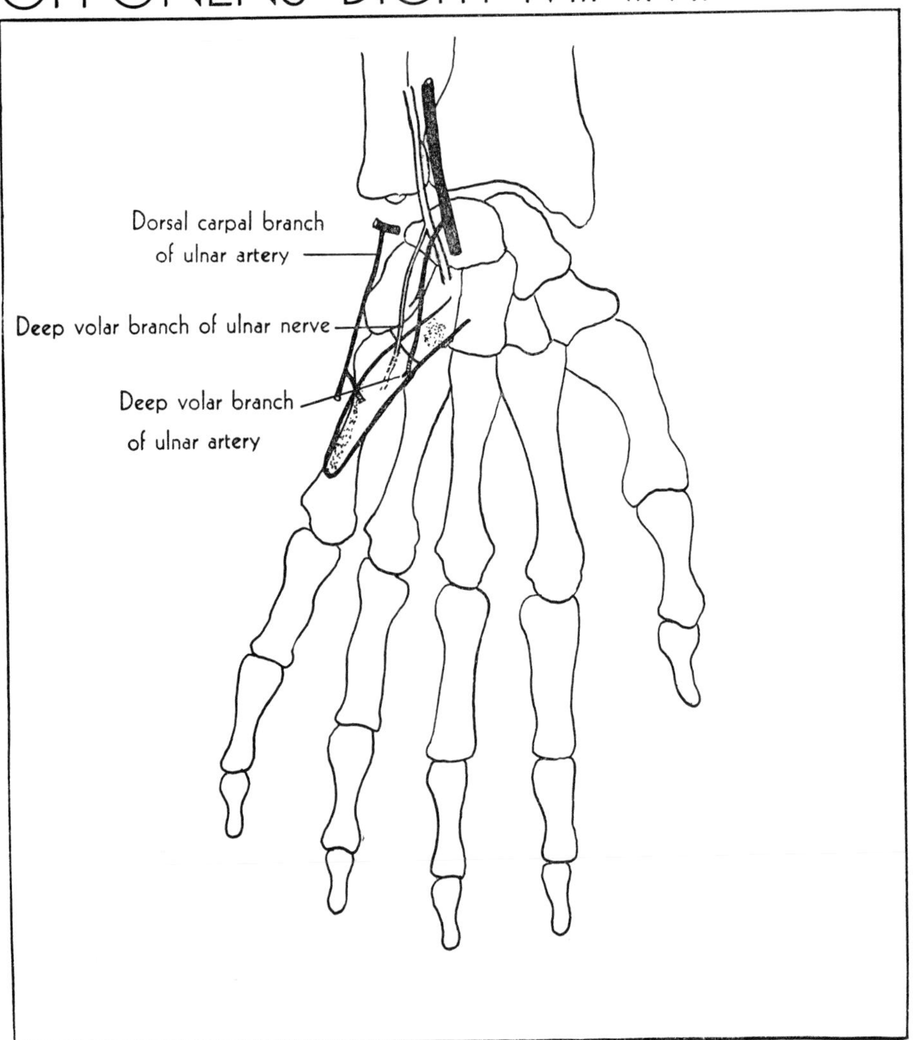

Dorsal carpal branch
of ulnar artery
Deep volar branch of ulnar nerve
Deep volar branch
of ulnar artery

ORIGIN: Transverse carpal ligament, hamulus of hamate bone

INSERTION: Ulnar margin of 5th metacarpal bone

FUNCTION: Draws 5th metacarpal bone forward to face thumb, deepens hollow of hand

NERVE: Deep volar branch of ulnar

ARTERY: Deep volar branch of ulnar, dorsal carpal branch of ulnar

REFERENCES:	GRAY	GRANT'S ATLAS
Muscle	485	61, 68, 81
Nerve	485, 961, 963, 975	68
Artery	624	71, 81

LUMBRICALES

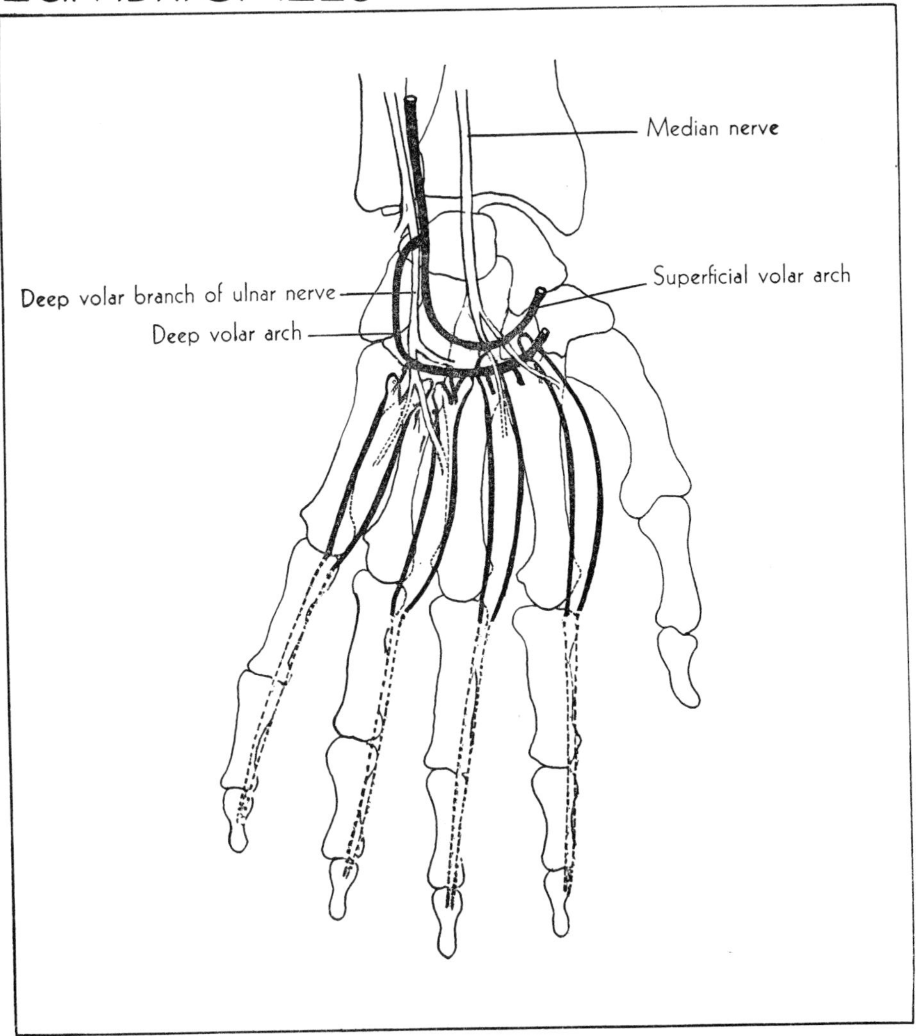

Median nerve
Superficial volar arch
Deep volar branch of ulnar nerve
Deep volar arch

ORIGIN: There are 4 lumbricales, all arising from tendons of flexor digitorum profundus: 1st from radial side of tendon for index finger, 2d from radial side of tendon for middle finger, 3d from adjacent sides of tendons for middle and ring fingers, 4th from adjacent sides of tendons for ring and little fingers

INSERTION: With tendons of extensor digitorum and interossei into bases of terminal phalanges of medial 4 digits

FUNCTION: Flex fingers at metacarpophalangeal joints, extend fingers at interphalangeal joints

NERVE: Median, to lateral two muscles, deep volar branch of ulnar to medial two muscles

ARTERY: Superficial and deep volar arches

REFERENCES:	GRAY	GRANT'S ATLAS
Muscle	485	61, 68
Nerve	485, 961, 963, 971, 975	68
Artery	620, 624	67

INTEROSSEI DORSALES

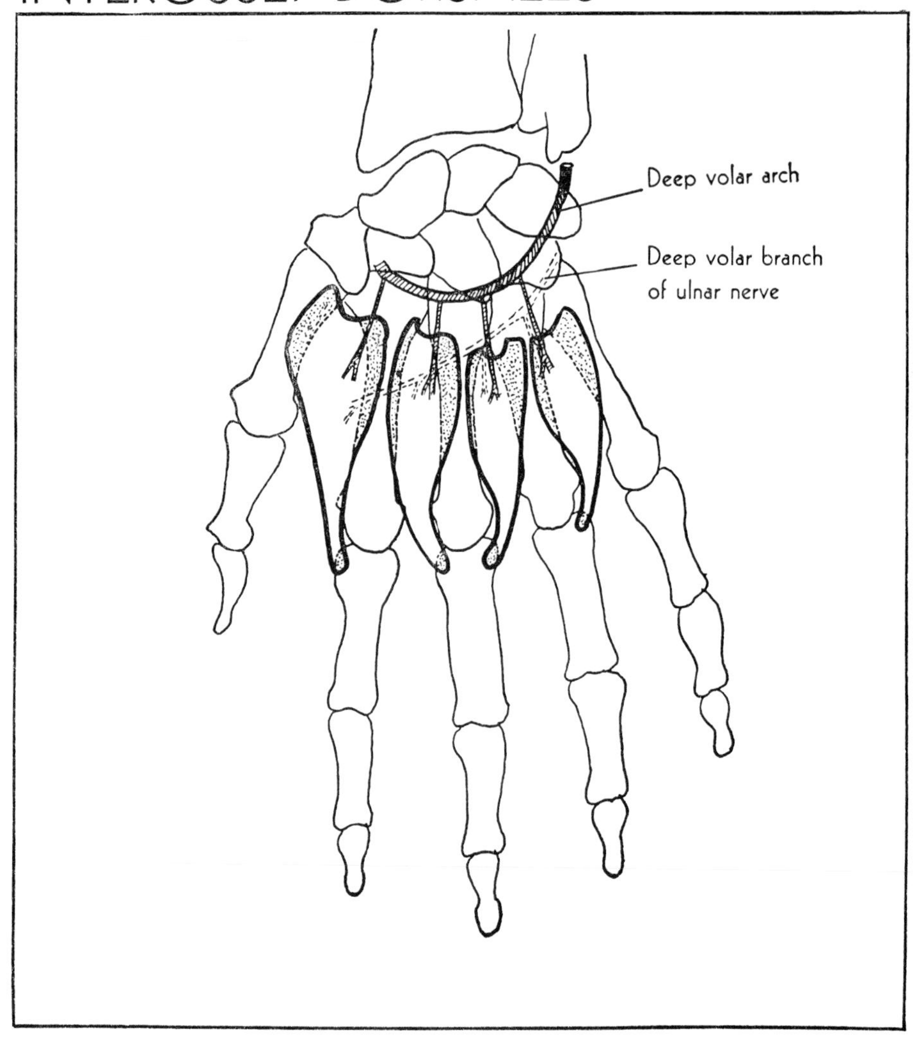

Deep volar arch
Deep volar branch of ulnar nerve

ORIGIN: There are 4 dorsal interossei; each arises by 2 heads from adjacent sides of metacarpal bones

INSERTION: 1st into radial side of proximal phalanx of 2d digit, 2d into radial side of proximal phalanx of 3d digit, 3d into ulnar side of proximal phalanx of 3d digit, 4th into ulnar side of proximal phalanx of 4th digit

FUNCTION: Abduct index, middle and ring fingers from the mid line of the hand

NERVE: Deep volar branch of ulnar

ARTERY: Deep volar arch

REFERENCES:	GRAY	GRANT'S ATLAS
Muscle	485	72, 78, 87
Nerve	486, 961, 963, 975	72
Artery	620	71.1, 75

INTEROSSEI PALMARES

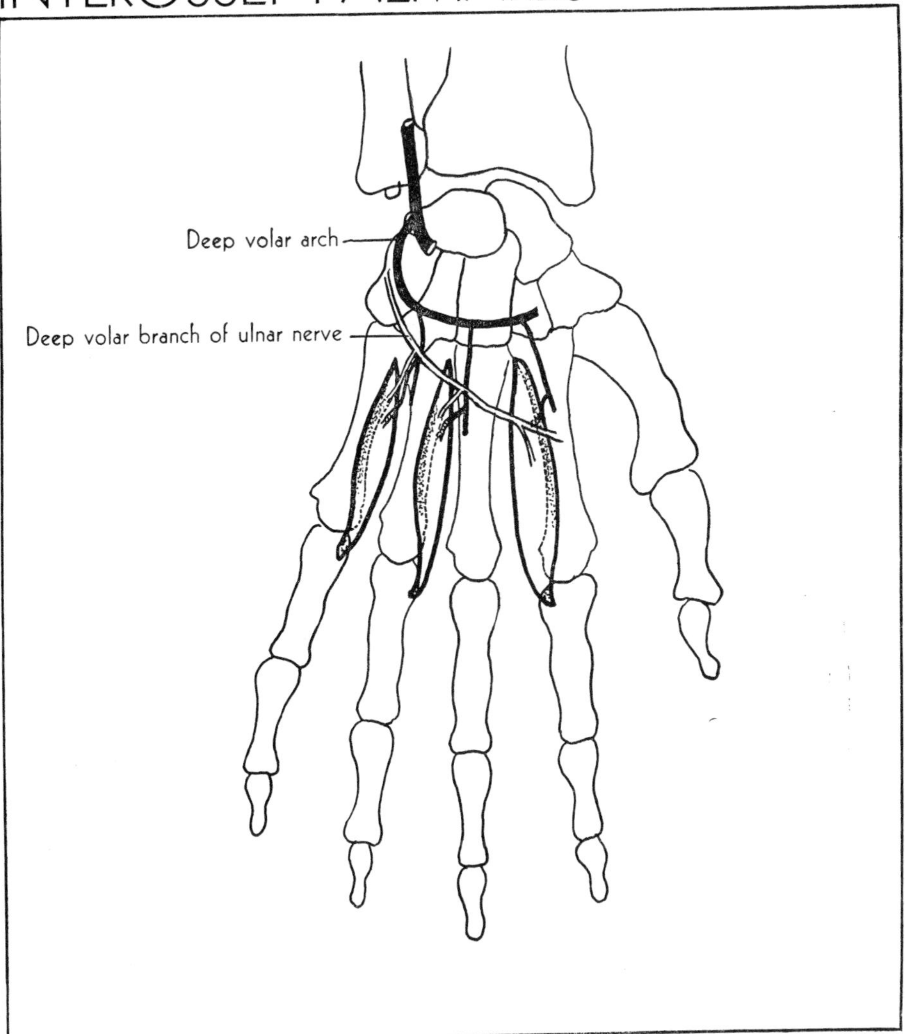

Deep volar arch
Deep volar branch of ulnar nerve

ORIGIN: There are 3 volar interossei: 1st from ulnar side of 2d metacarpal bone, 2d from radial side of 4th metacarpal bone, 3d from radial side of 5th metacarpal bone

INSERTION: 1st into ulnar side of proximal phalanx of 2d digit, 2d into radial side of proximal phalanx of 4th digit, 3d into radial side of proximal phalanx of 5th digit

FUNCTION: Each muscle adducts digit into which it is inserted toward middle digit

NERVE: Deep volar branch of ulnar

ARTERY: Deep volar arch

REFERENCES:	GRAY	GRANT'S ATLAS
Muscle	486	72
Nerve	487, 961, 963, 975	72
Artery	620	71.1

PSOAS MAJOR

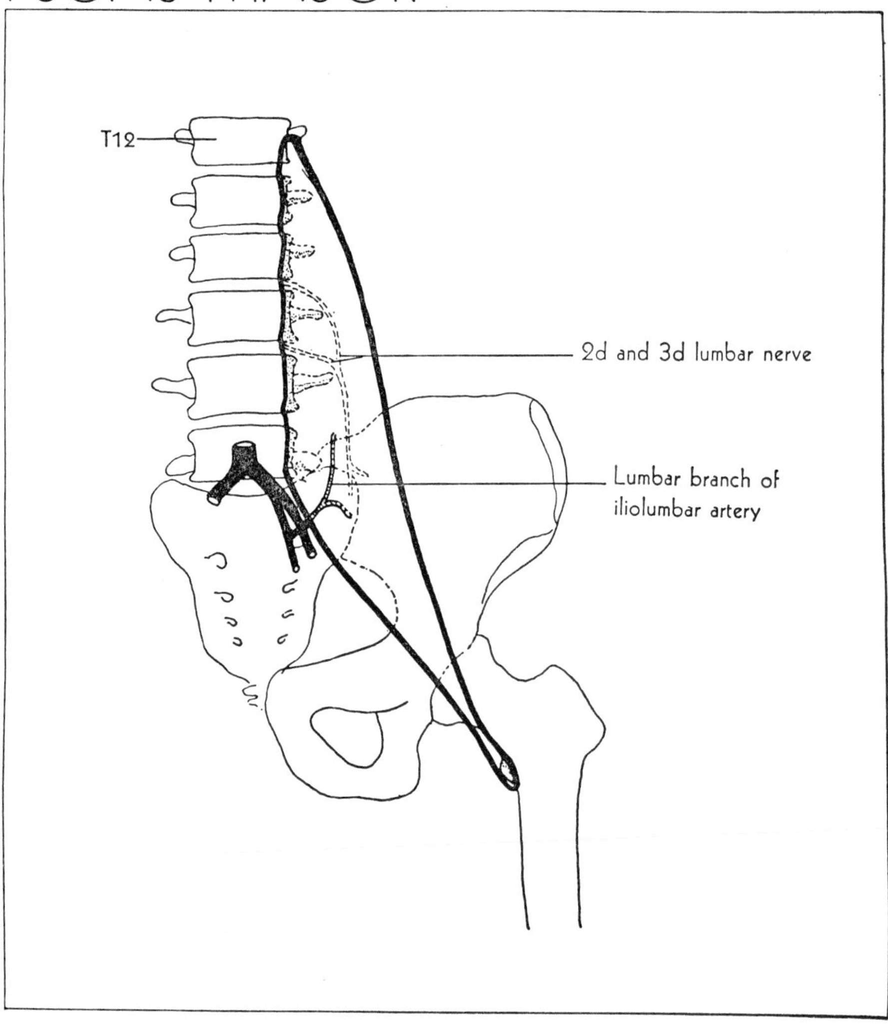

T12
2d and 3d lumbar nerve
Lumbar branch of iliolumbar artery

ORIGIN: Anterior surface of transverse processes and bodies of lumbar vertebrae, corresponding intervertebral discs

INSERTION: Lesser trochanter of femur (with iliacus forms iliopsoas)

FUNCTION: Flexes thigh, flexes vertebral column on pelvis when leg is fixed

NERVE: 2d and 3d lumbar

ARTERY: Lumbar branch of iliolumbar

REFERENCES: GRAY		GRANT'S ATLAS
Muscle	489	179, 190, 256
Nerve	489, 983	Not shown
Artery	649	Not shown

PSOAS MINOR

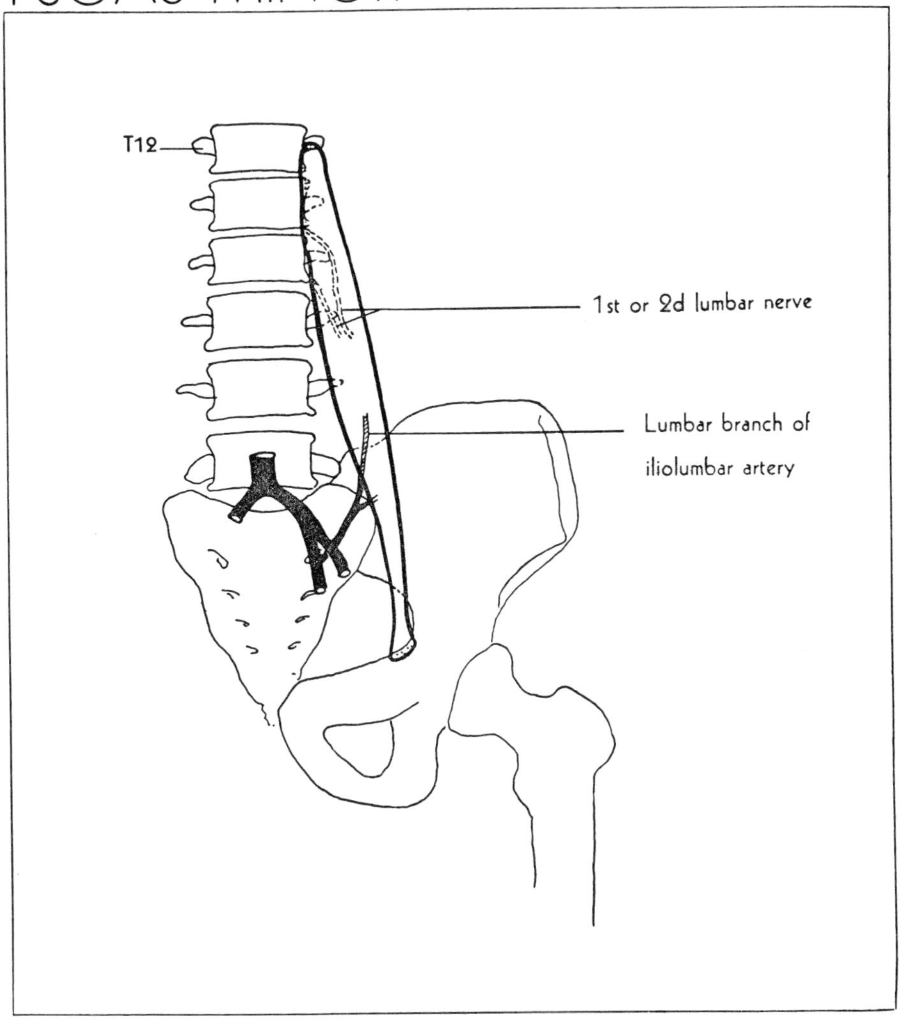

T12
1st or 2d lumbar nerve
Lumbar branch of
iliolumbar artery

ORIGIN: Vertebral margins of 12th thoracic and 1st lumbar vertebra, corresponding intervertebral disc

INSERTION: Pectineal line, iliopectineal eminence

FUNCTION: Flexes pelvis on vertebral column, assists psoas major in flexing vertebral column on pelvis. This muscle is inconstant; absent in 40 per cent of bodies

NERVE: 1st or 2d lumbar

ARTERY: Lumbar branch of iliolumbar

REFERENCES:	GRAY	GRANT'S ATLAS
Muscle	489	260
Nerve	489, 983	Not shown
Artery	649	Not shown

ILIACUS

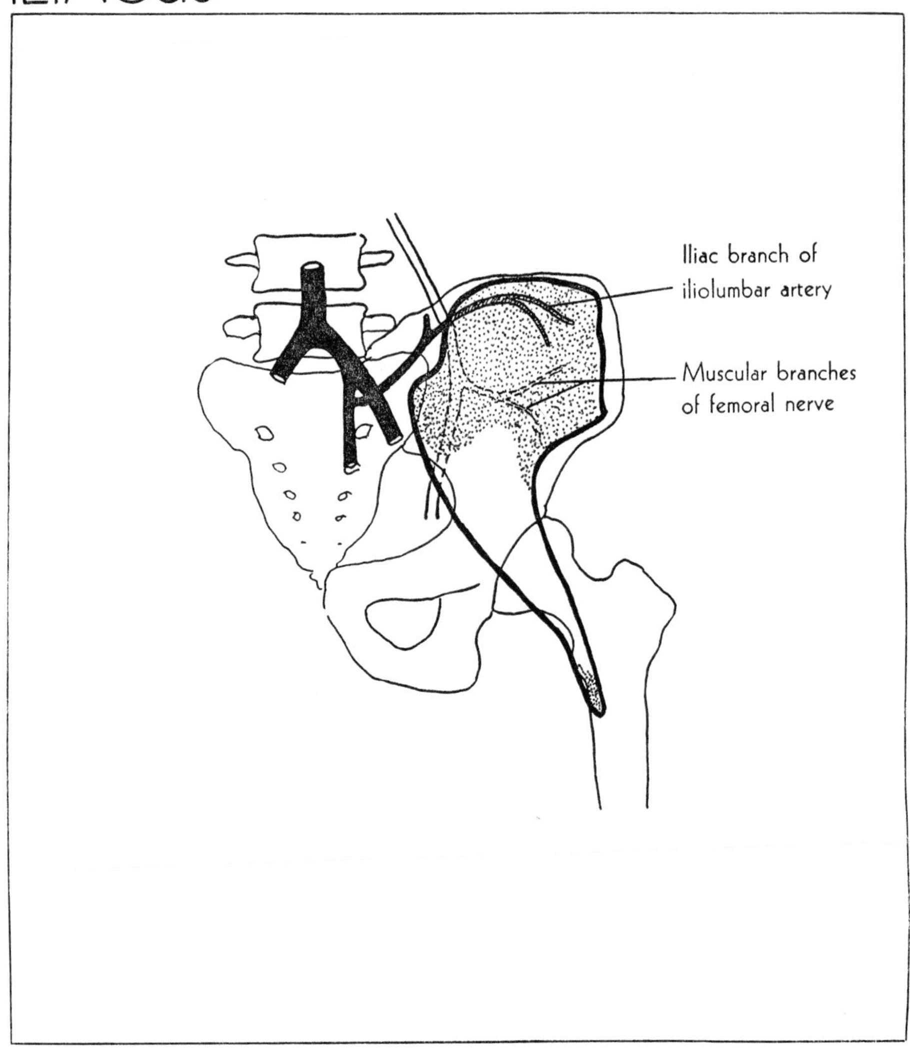

Iliac branch of iliolumbar artery
Muscular branches of femoral nerve

ORIGIN: Upper two-thirds of iliac fossa; iliac crest; anterior sacroiliac, lumbosacral, and iliolumbar ligaments; ala of sacrum

INSERTION: Tendon of psoas major, lesser trochanter, capsule of hip joint, body of femur (with psoas major forms iliopsoas)

FUNCTION: Flexes thigh, tilts pelvis forward when leg is fixed

NERVE: Muscular branches of femoral

ARTERY: Iliac branch of iliolumbar

REFERENCES:	GRAY	GRANT'S ATLAS
Muscle	489	261
Nerve	489, 983-985, 989	190
Artery	649	190

SARTORIUS

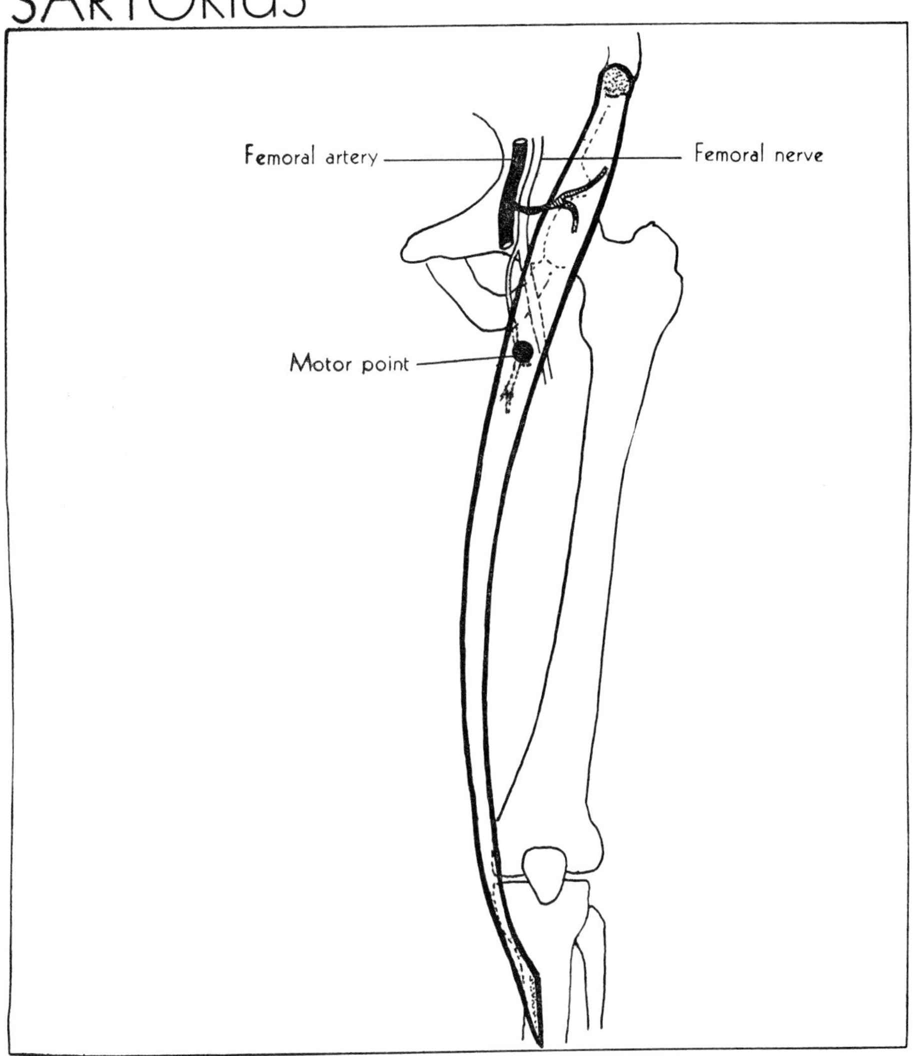
Femoral artery
Femoral nerve
Motor point

ORIGIN:	Anterior superior iliac spine, upper half of iliac notch
INSERTION:	Upper part of medial surface of tibia
FUNCTION:	Flexes leg on thigh, flexes thigh on pelvis, rotates thigh laterally
NERVE:	Muscular branches of femoral
ARTERY:	Muscular branches of femoral

REFERENCES: GRAY	GRANT'S ATLAS
Muscle 492	258, 260
Nerve 493, 983-985, 989	263
Artery 658	Not shown

RECTUS FEMORIS

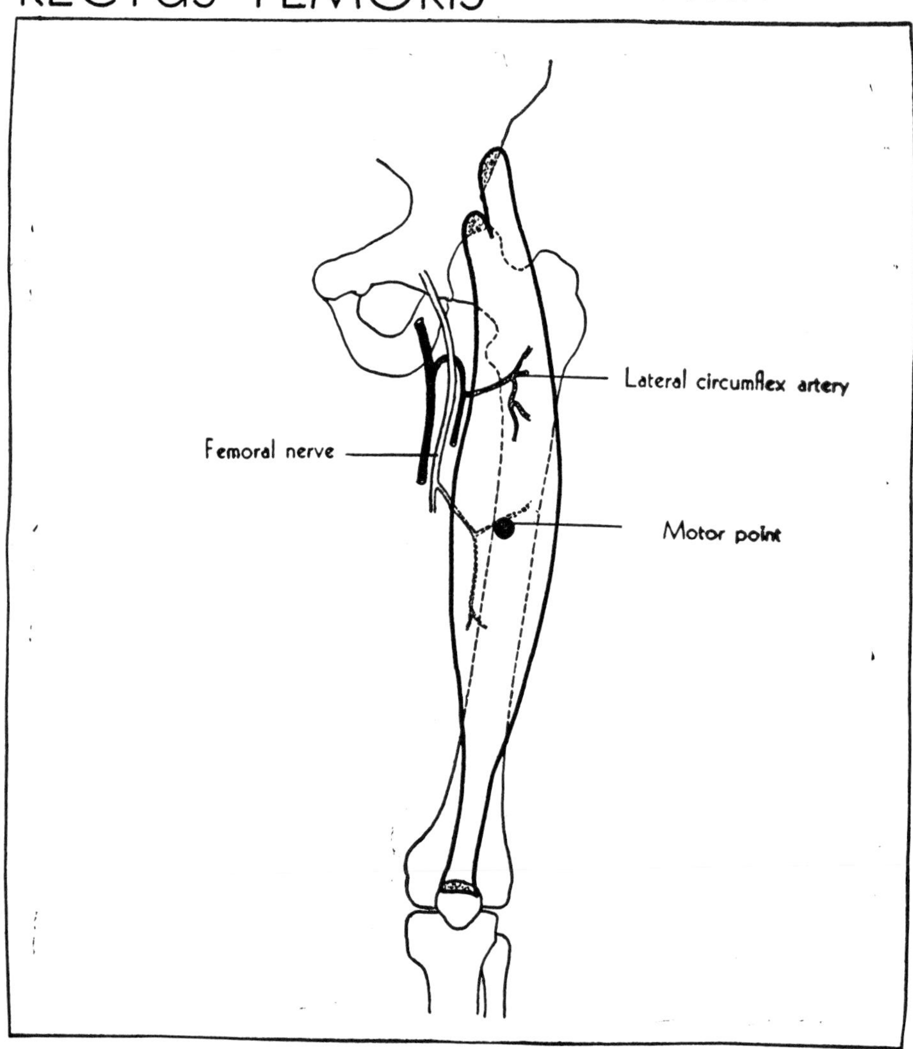
Lateral circumflex artery
Femoral nerve
Motor point

ORIGIN: (Rectus femoris is 1 of 4 muscles comprising quadriceps femoris.) Straight head from anterior inferior iliac spine, reflected head from groove on upper brim of acetabulum

INSERTION: Upper border of patella; by ligamentum patellae into tibial tuberosity

FUNCTION: Extends leg, flexes thigh

NERVE: Muscular branches of femoral

ARTERY: Lateral femoral circumflex

REFERENCES: GRAY		GRANT'S ATLAS
Muscle	493-495	260
Nerve	494, 983-985, 991	263
Artery	658	263

VASTUS LATERALIS

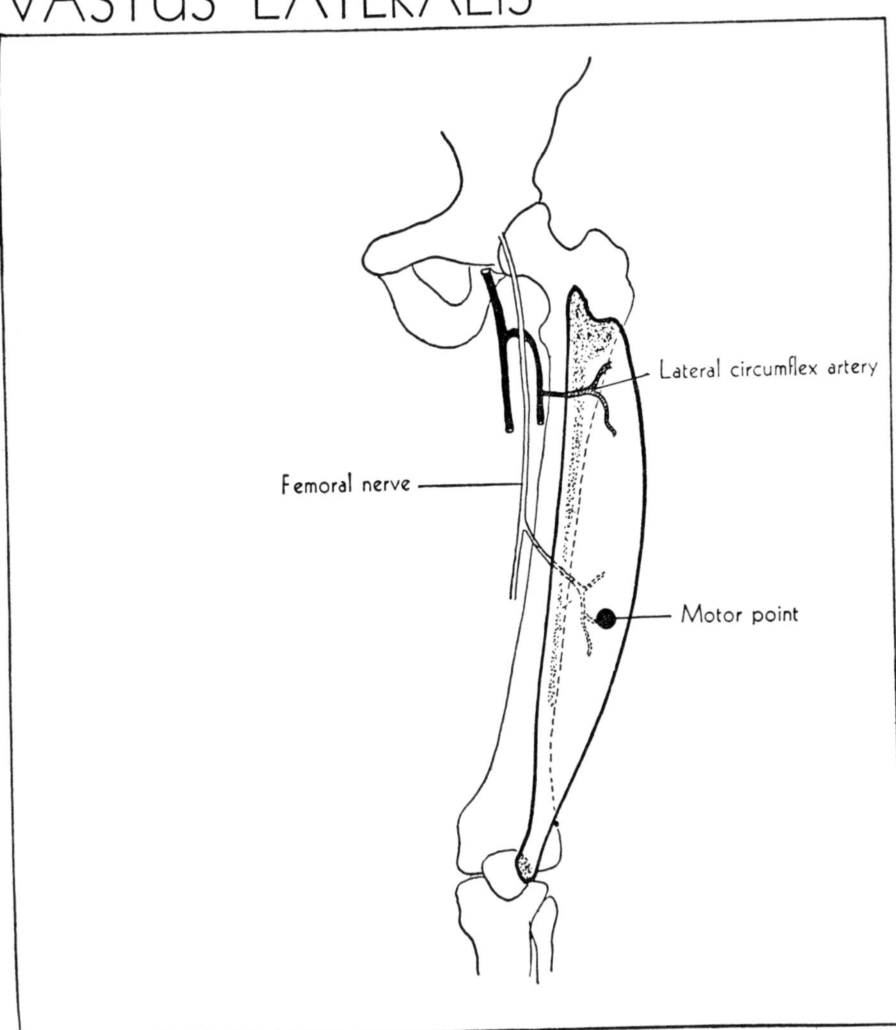
Lateral circumflex artery
Femoral nerve
Motor point

ORIGIN: (Vastus lateralis is 1 of 4 muscles comprising quadriceps femoris.) Capsule of hip joint, intertrochanteric line, greater trochanter, gluteal tuberosity, linea aspera, lateral intermuscular septum

INSERTION: Lateral border by patella, by ligamentum patellae into tibial tuberosity

FUNCTION: Extends leg

NERVE: Muscular branches of femoral

ARTERY: Latera femoral circumflex

REFERENCES: GRAY		GRANT'S ATLAS
Muscle	493-495	260
Nerve	494, 983-985, 991	263
Artery	658	263

VASTUS MEDIALIS

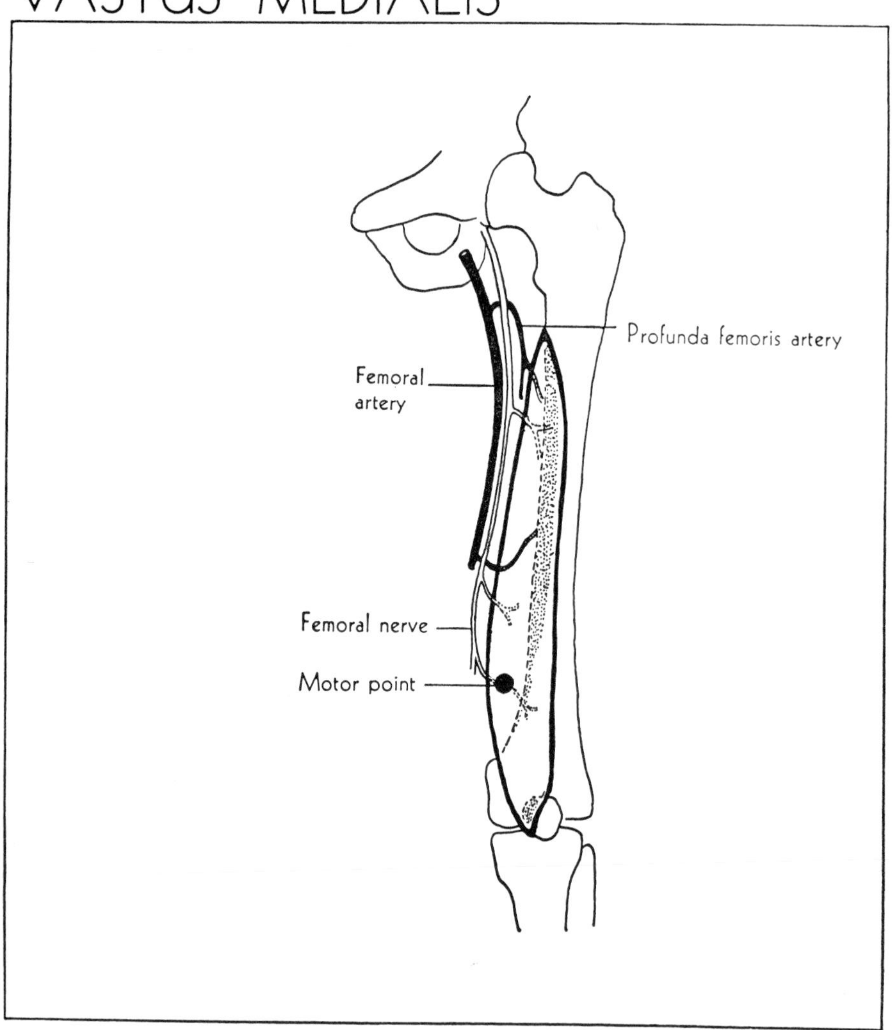

Profunda femoris artery
Femoral artery
Femoral nerve
Motor point

ORIGIN: (Vastus medialis is 1 of 4 muscles comprising quadriceps femoris.) Lower half of intertrochanteric line, linea aspera, medial supracondylar line, medial intermuscular septum, tendon of adductor magnus

INSERTION: Quadriceps femoris tendon, medial border of patella, capsule of knee joint, by ligamentum patellae into tibial tuberosity

FUNCTION: Extends leg and draws patella medially

NERVE: Muscular branches of femoral

ARTERY: Muscular branches of femoral, muscular branches of profunda femoris genicular branches of popliteal

REFERENCES:	GRAY	GRANT'S ATLAS
Muscle	493-495	260
Nerve	494, 983-985, 991	263
Artery	658, 660, 663-664	Not shown

VASTUS INTERMEDIUS *

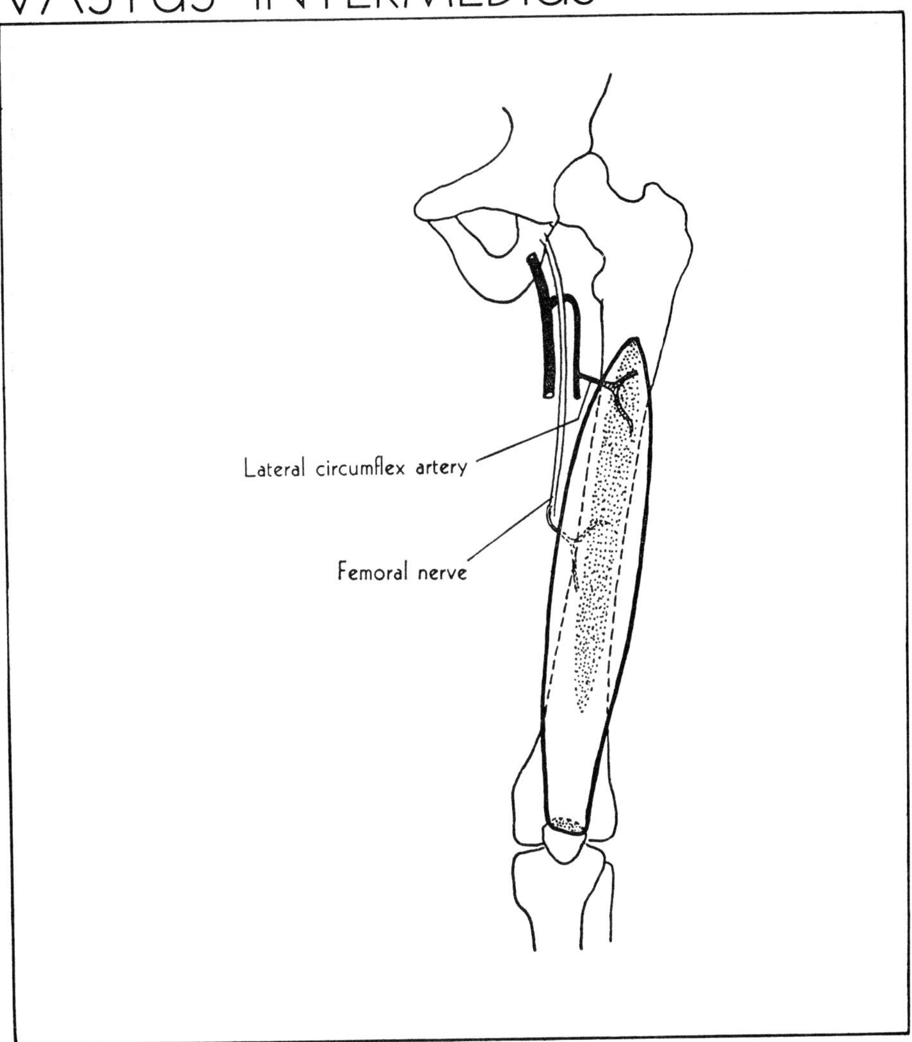

Lateral circumflex artery
Femoral nerve

ORIGIN: (Vastus intermedius is 1 of 4 muscles comprising quadriceps femoris.) Upper two-thirds of anterior and lateral surface of femur, lower half of linea aspera, upper part of lateral supracondylar line, lateral intermuscular septum

INSERTION: Deep surface of tendons of rectus and vasti muscles, by ligamentum patellae into tibial tuberosity

FUNCTION: Extends leg

NERVE: Muscular branches of femoral

ARTERY: Lateral femoral circumflex

REFERENCES:	GRAY	GRANT'S ATLAS
Muscle	493-495	261
Nerve	494, 495, 983-985, 991	263
Artery	658	263

***Articularis genus: a few separate muscle bundles arising deep to V. intermedius; tenses** capsule of knee joint.

GRACILIS

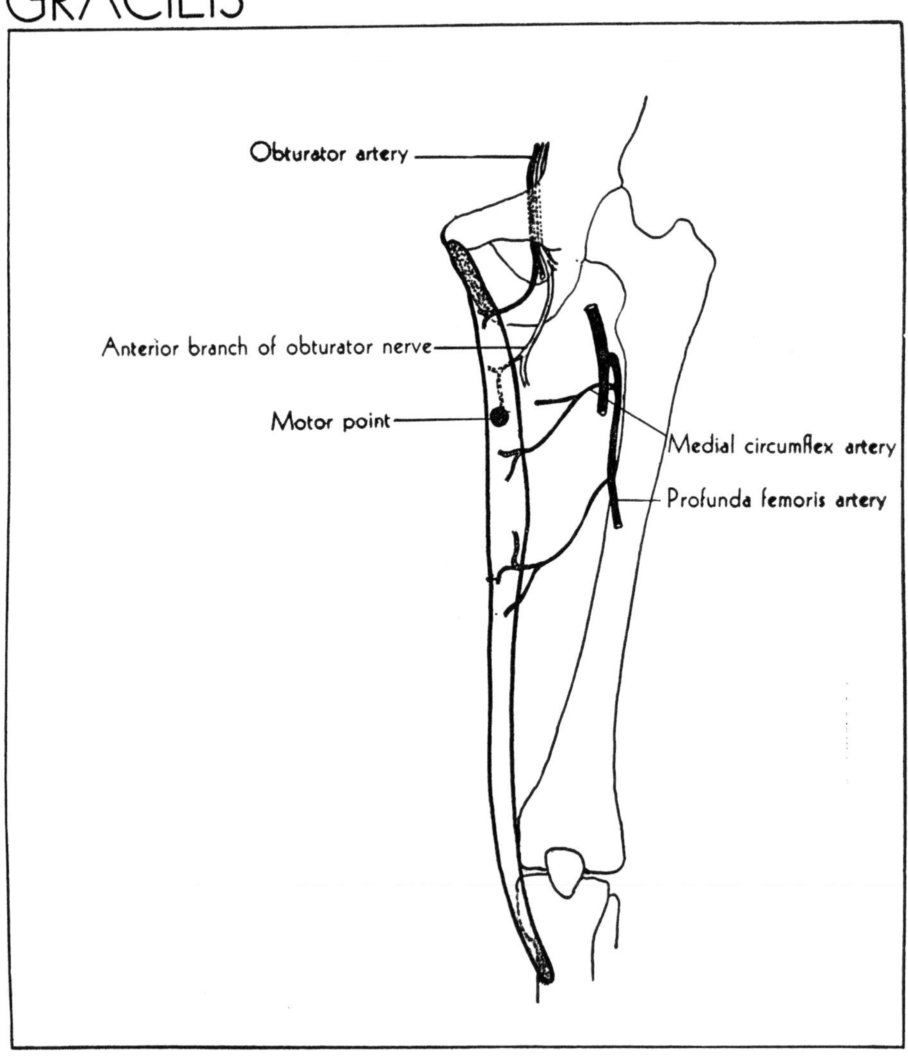

Obturator artery
Anterior branch of obturator nerve
Motor point
Medial circumflex artery
Profunda femoris artery

ORIGIN: Lower half of pubic symphysis, upper half of pubic arch
INSERTION: Upper part of medial surface of tibia
FUNCTION: Flexes and medially rotates leg, adducts thigh
NERVE: Anterior branch of obturator
ARTERY: Muscular branches of profunda femoris, obturator, medial femoral circumflex

REFERENCES:	GRAY	GRANT'S ATLAS
Muscle	495	262
Nerve	495, 983-985, 987	263
Artery	646, 658, 660	263

PECTINEUS

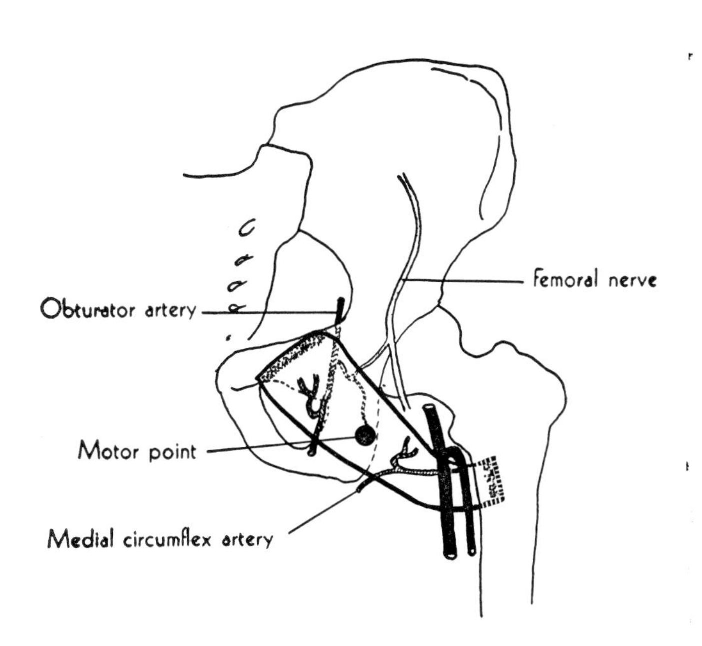
Femoral nerve
Obturator artery
Motor point
Medial circumflex artery

ORIGIN: Pectineal line, surface of pubis between iliopectineal eminence and pubic tubercle

INSERTION: Line extending from lesser trochanter to linea aspera

FUNCTION: Adducts, flexes, medially rotates thigh

NERVE: Muscular branches of femoral

ARTERY: Medial femoral circumflex, obturator

REFERENCES:	GRAY	GRANT'S ATLAS
Muscle	495	259
Nerve	495, 983-985, 989	263
Artery	646, 658	263

ADDUCTOR LONGUS

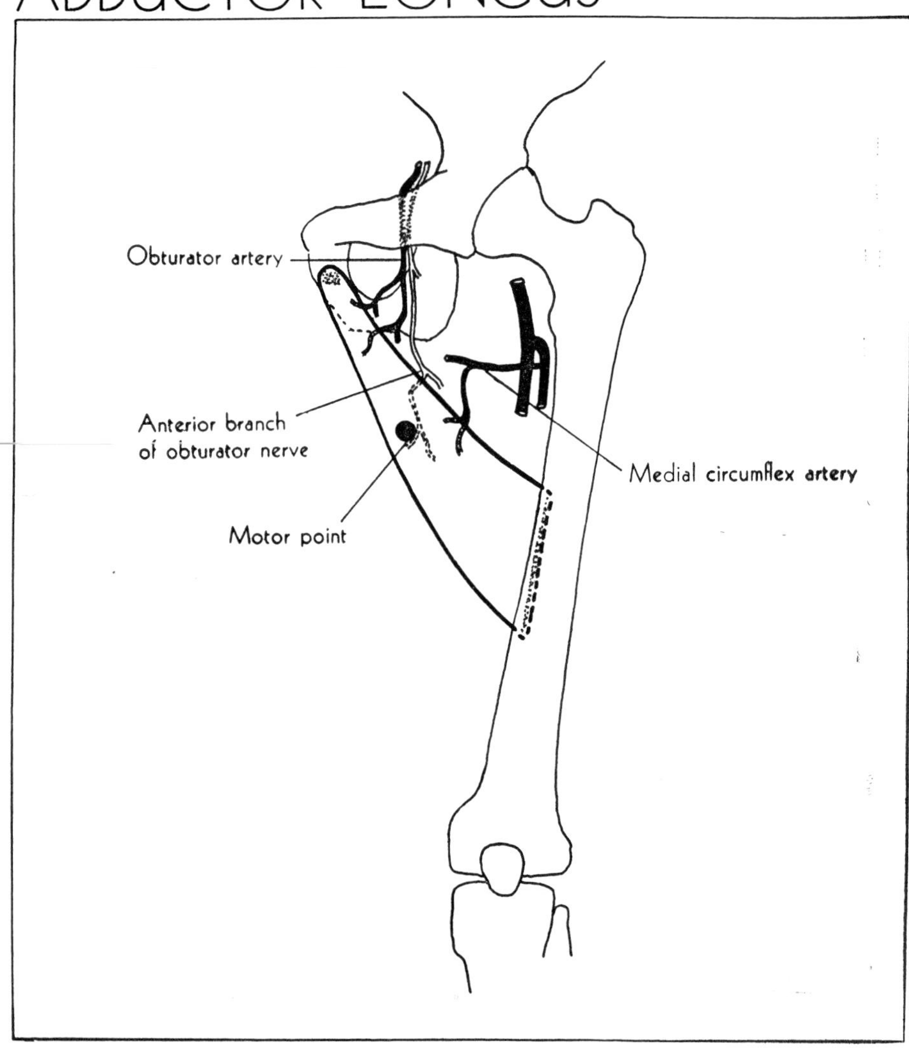

Obturator artery
Anterior branch
of obturator nerve
Motor point
Medial circumflex artery

ORIGIN:	Front of pubis in angle between crest and symphysis
INSERTION:	Middle half of medial lip of linea aspera
FUNCTION:	Adducts thigh, and assists in flexing it. Rotator action is controversial
NERVE:	Anterior branch of obturator
ARTERY:	Medial femoral circumflex, obturator

REFERENCES:	GRAY	GRANT'S ATLAS
Muscle	495	259
Nerve	495, 983-985, 987	263
Artery	646, 658	263

ADDUCTOR BREVIS

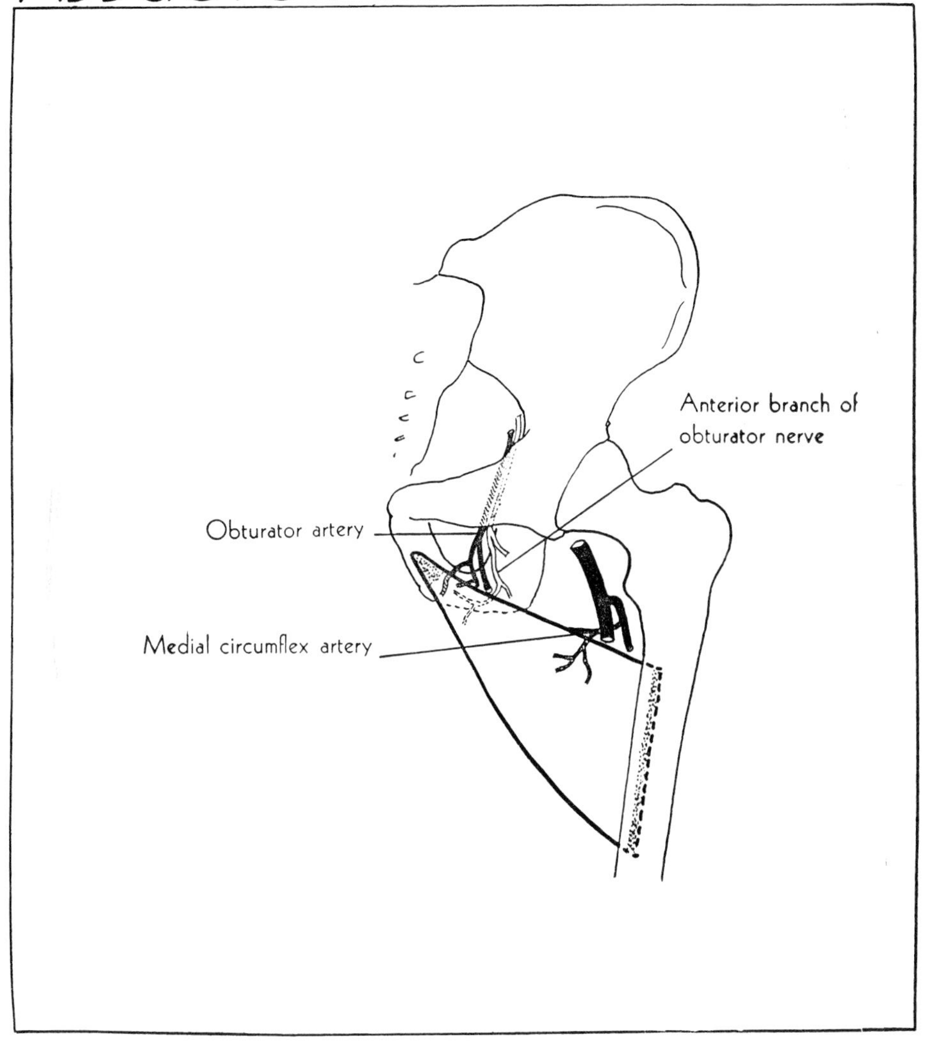
Anterior branch of
obturator nerve
Obturator artery
Medial circumflex artery

ORIGIN: Outer surface of inferior ramus of pubis
INSERTION: Line extending from lesser trochanter to linea aspera
FUNCTION: Adducts thigh, and assists in flexing it. Rotator action is controversial
NERVE: Anterior branch of obturator
ARTERY: Medial femoral circumflex, obturator

REFERENCES:	GRAY	GRANT'S ATLAS
Muscle	495	261
Nerve	496, 983-985, 987	Not shown
Artery	646, 658	263

ADDUCTOR MAGNUS

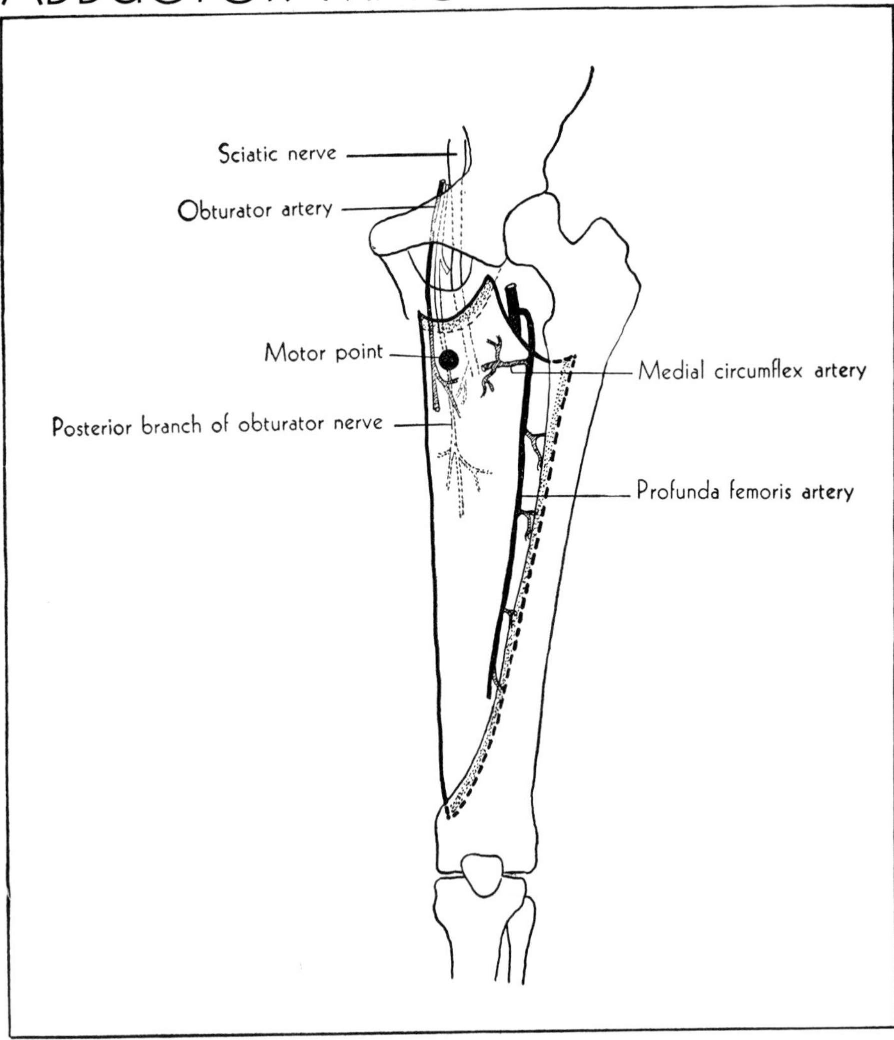
Sciatic nerve
Obturator artery
Motor point
Posterior branch of obturator nerve
Medial circumflex artery
Profunda femoris artery

ORIGIN: Ischial tuberosity, rami of ischium and pubis

INSERTION: Line extending from greater trochanter to linea aspera, linea aspera, medial supracondylar line, adductor tubercle

FUNCTION: Adducts thigh; upper portion flexes it; lower portion extends it. Rotating action is controversial

NERVE: Posterior branch of obturator, sciatic

ARTERY: Medial femoral circumflex, perforating branches of profunda femoris, obturator, muscular branches of popliteal

REFERENCES:	GRAY	GRANT'S ATLAS
Muscle	496	269
Nerve	497, 983-985, 987, 997	271
Artery	646, 658, 660, 662	263, 271

GLUTEUS MAXIMUS

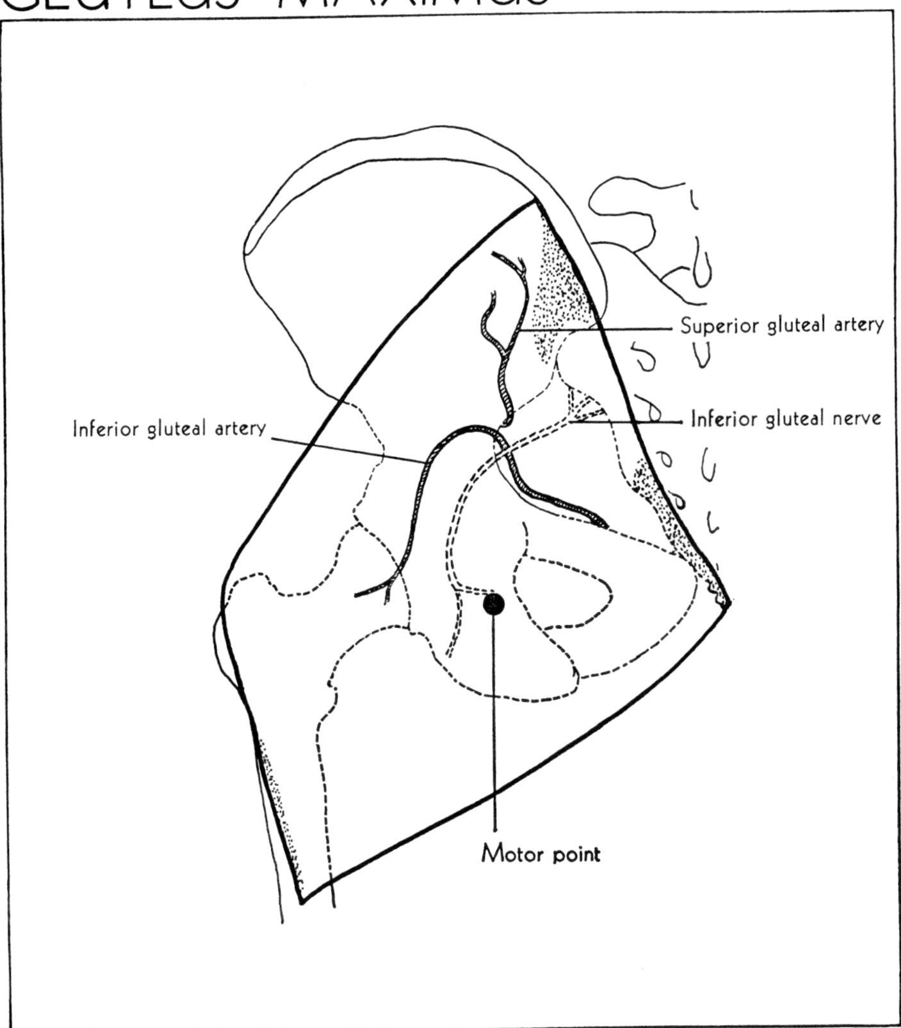

Superior gluteal artery
Inferior gluteal artery
Inferior gluteal nerve
Motor point

ORIGIN: Posterior gluteal line, tendon of sacrospinalis, dorsal surface of sacrum and coccyx, sacrotuberous ligament

INSERTION: Gluteal tuberosity of femur, iliotibial tract

FUNCTION: Extends thigh, assists in adducting and laterally rotating it; acting on insertion, muscle extends trunk

NERVE: Inferior gluteal

ARTERY: Superior gluteal, inferior gluteal, 1st perforating branch of profunda femoris

REFERENCES:	GRAY	GRANT'S ATLAS
Muscle	497	267
Nerve	497, 993-994	270, 271
Artery	649, 660	270, 271

GLUTEUS MEDIUS

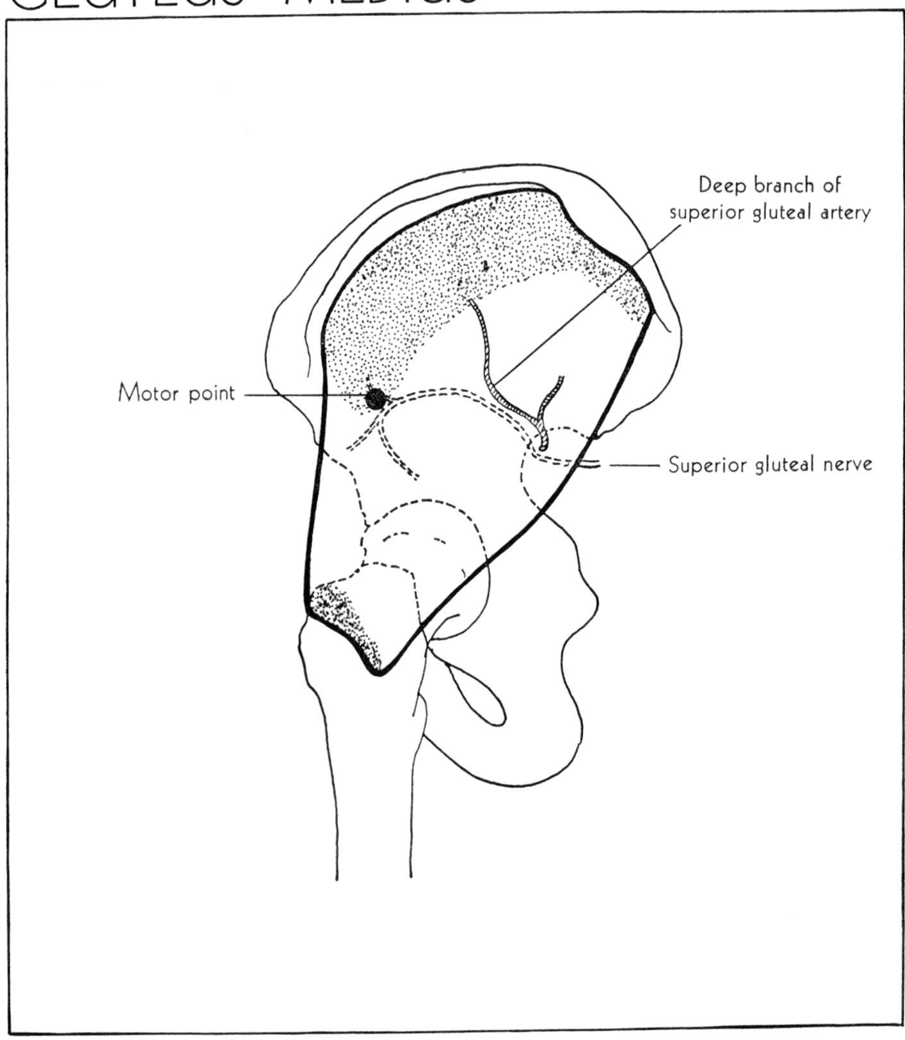

Deep branch of superior gluteal artery
Motor point
Superior gluteal nerve

ORIGIN: Outer surface of ilium from iliac crest and posterior gluteal line above, to anterior gluteal line below, gluteal aponeurosis

INSERTION: Lateral surface of greater trochanter

FUNCTION: Abducts thigh, rotates thigh medially when limb is extended

NERVE: Superior gluteal

ARTERY: Deep branch of superior gluteal

REFERENCES:	GRAY	GRANT'S ATLAS
Muscle	497	268
Nerve	497, 993-994	271
Artery	649	270

GLUTEUS MINIMUS

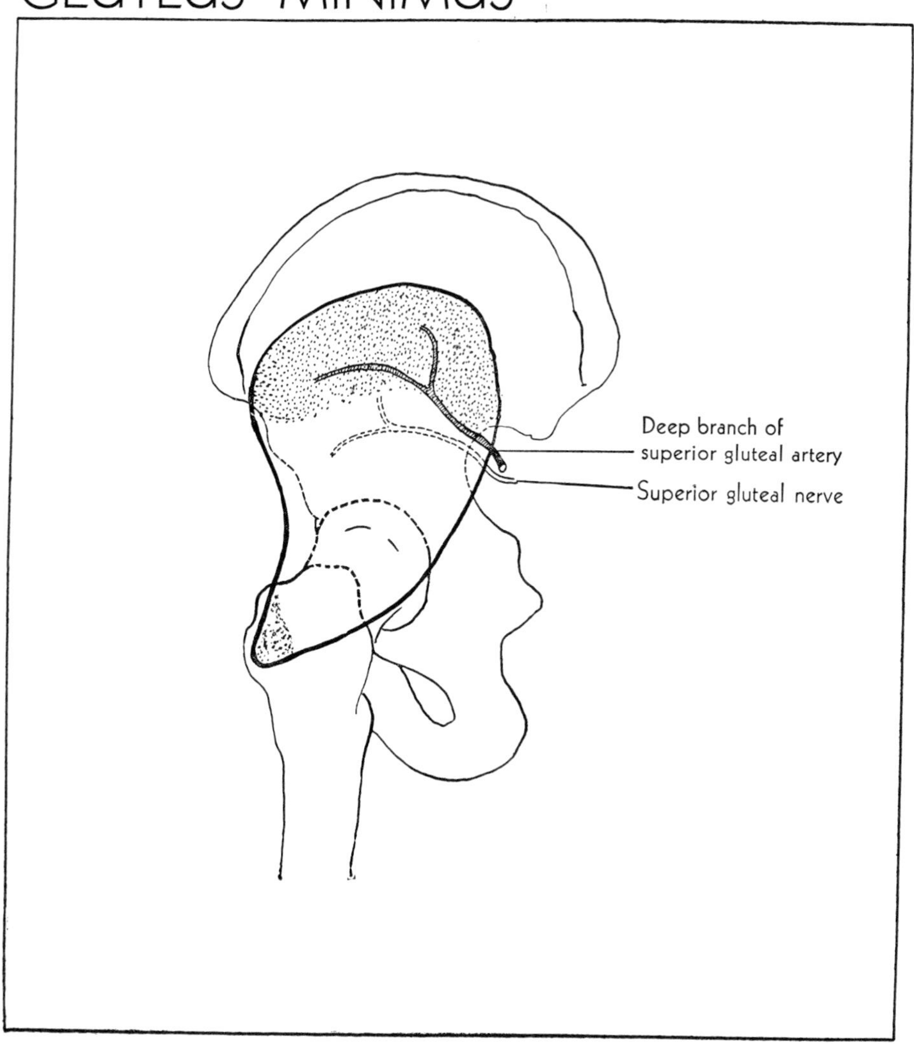
Deep branch of
superior gluteal artery
Superior gluteal nerve

ORIGIN: Outer surface of ilium between anterior and inferior gluteal lines margin of greater sciatic notch

INSERTION: Anterior border of greater trochanter

FUNCTION: Abducts thigh, rotates thigh medially when limb is extended

NERVE: Superior gluteal

ARTERY: Deep branch of superior gluteal

REFERENCES: GRAY	GRANT'S ATLAS
Muscle 497	271
Nerve 498, 993-994	271
Artery 649	271

TENSOR FASCIAE LATAE

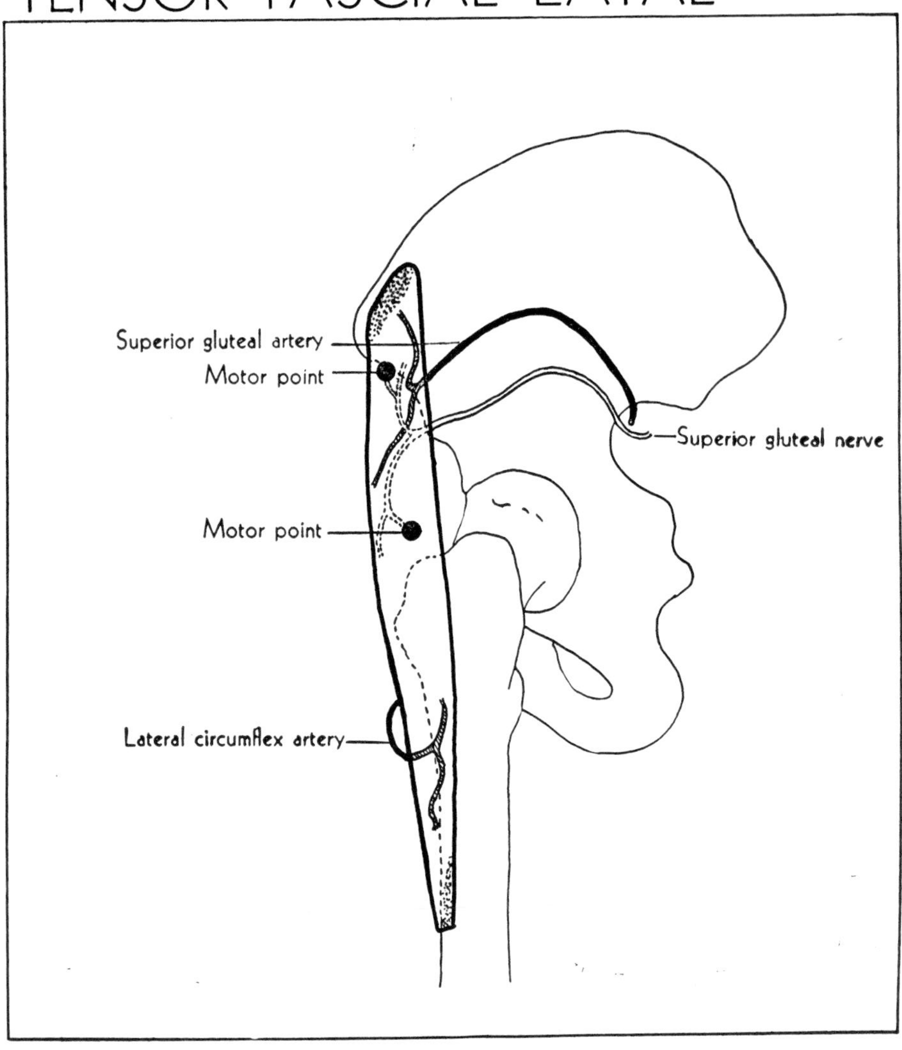

Superior gluteal artery
Motor point
Superior gluteal nerve
Motor point
Lateral circumflex artery

ORIGIN:	Anterior part of outer lip of iliac crest, anterior border of ilium
INSERTION:	Middle third of thigh along iliotibial tract
FUNCTION:	Tenses fascia lata counteracting backward pull of gluteus maximus on iliotibial tract; assists in flexing, abducting, and medially rotating thigh
NERVE:	Superior gluteal
ARTERY:	Lateral femoral circumflex, superior gluteal

REFERENCES:	GRAY	GRANT'S ATLAS
Muscle	498	259
Nerve	499, 993-994	Not shown
Artery	649, 658	259

PIRIFORMIS

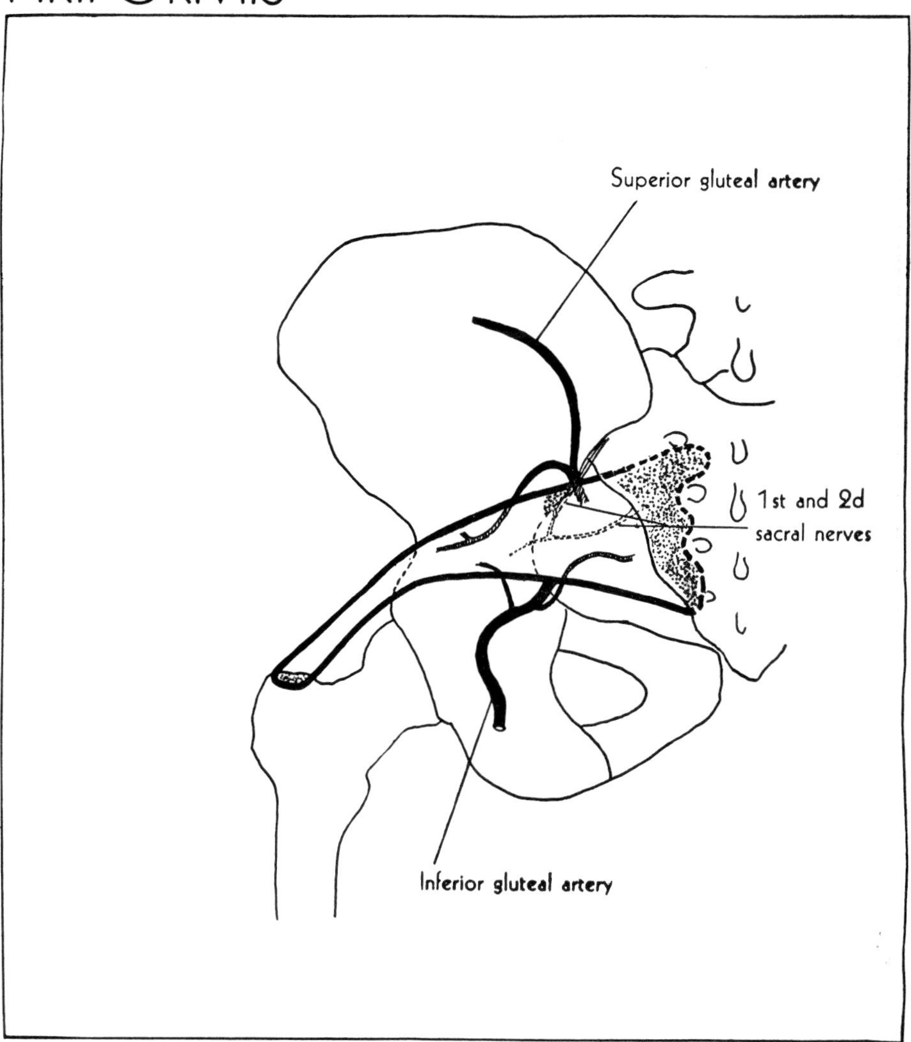

Superior gluteal artery
1st and 2d
sacral nerves
Inferior gluteal artery

ORIGIN: Pelvic surface of sacrum between anterior sacral foramina, margin of greater sciatic foramen, sacrotuberous ligament

INSERTION: Upper border of greater trochanter of femur

FUNCTION: Rotates thigh laterally, abducts thigh when limb is flexed

NERVE: 1st and 2d sacral

ARTERY: Superior gluteal, inferior gluteal, internal pudendal

REFERENCES:	GRAY	GRANT'S ATLAS
Muscle	499	273
Nerve	499, 993-994	213
Artery	648, 649	270, 271

OBTURATOR INTERNUS

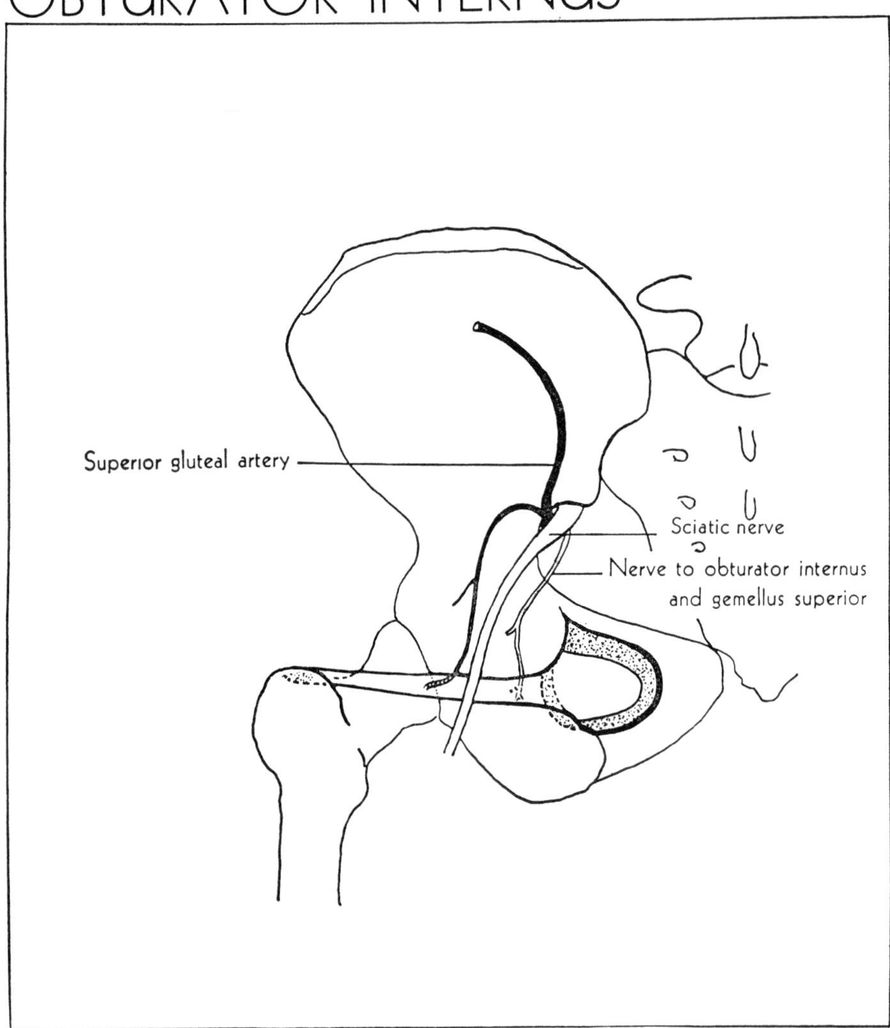
Superior gluteal artery
Sciatic nerve
Nerve to obturator internus and gemellus superior

ORIGIN: Margins of obturator foramen, obturator membrane, pelvic surface of hip bone behind and above obturator foramen, obturator fascia

INSERTION: Medial surface of greater trochanter

FUNCTION: Rotates thigh laterally, abducts thigh when limb is flexed

NERVE: Nerve to obturator internus and gemellus superior

ARTERY: Muscular branches of internal pudendal; superior gluteal

REFERENCES:	GRAY	GRANT'S ATLAS
Muscle	500	216
Nerve	501, 993-994	270
Artery	648, 649	216

GEMELLUS SUPERIOR

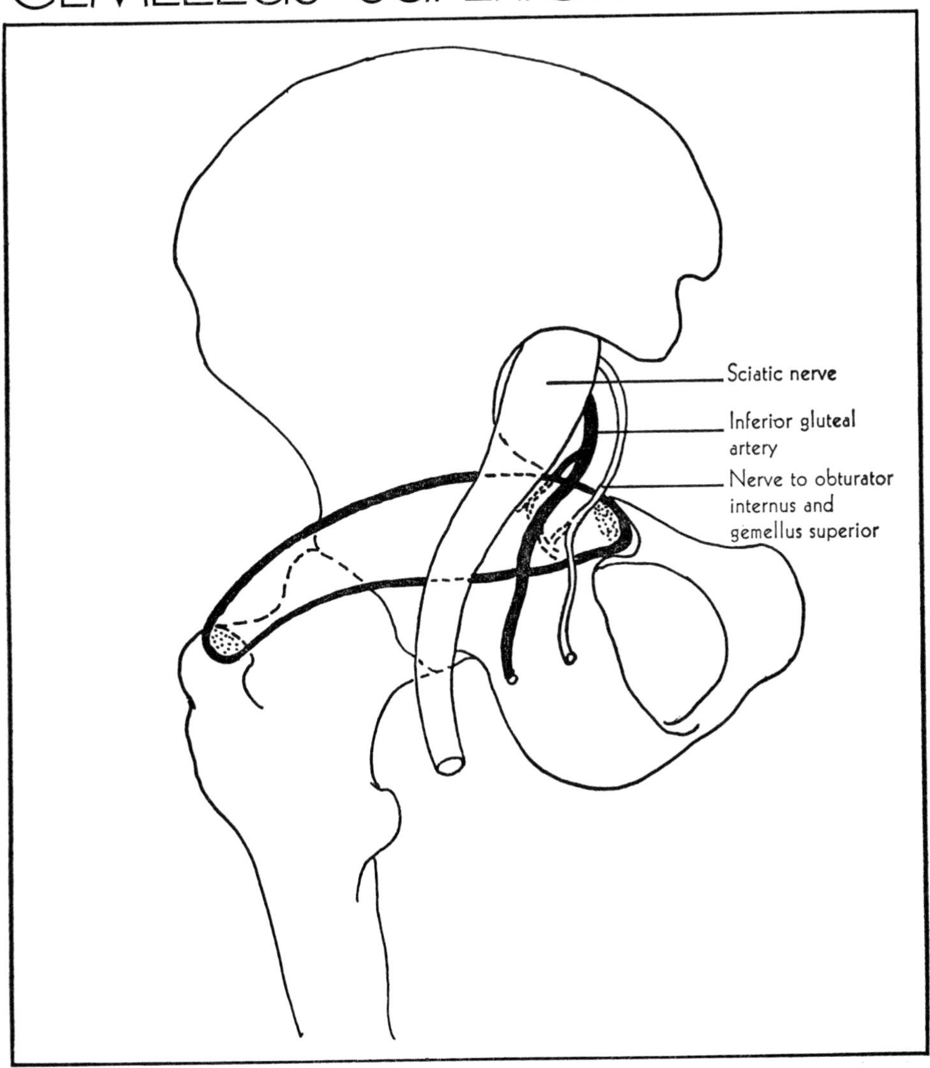
Sciatic nerve
Inferior gluteal artery
Nerve to obturator internus and gemellus superior

ORIGIN: Outer surface of ischial spine
INSERTION: Medial surface of greater trochanter, blends with obturator internus tendon
FUNCTION: Rotates thigh laterally
NERVE: Nerve to obturator internus and gemellus superior
ARTERY: Inferior gluteal

REFERENCES:	GRAY	GRANT'S ATLAS
Muscle	501	272
Nerve	501, 993-994	270, 271
Artery	649	270, 271

GEMELLUS INFERIOR

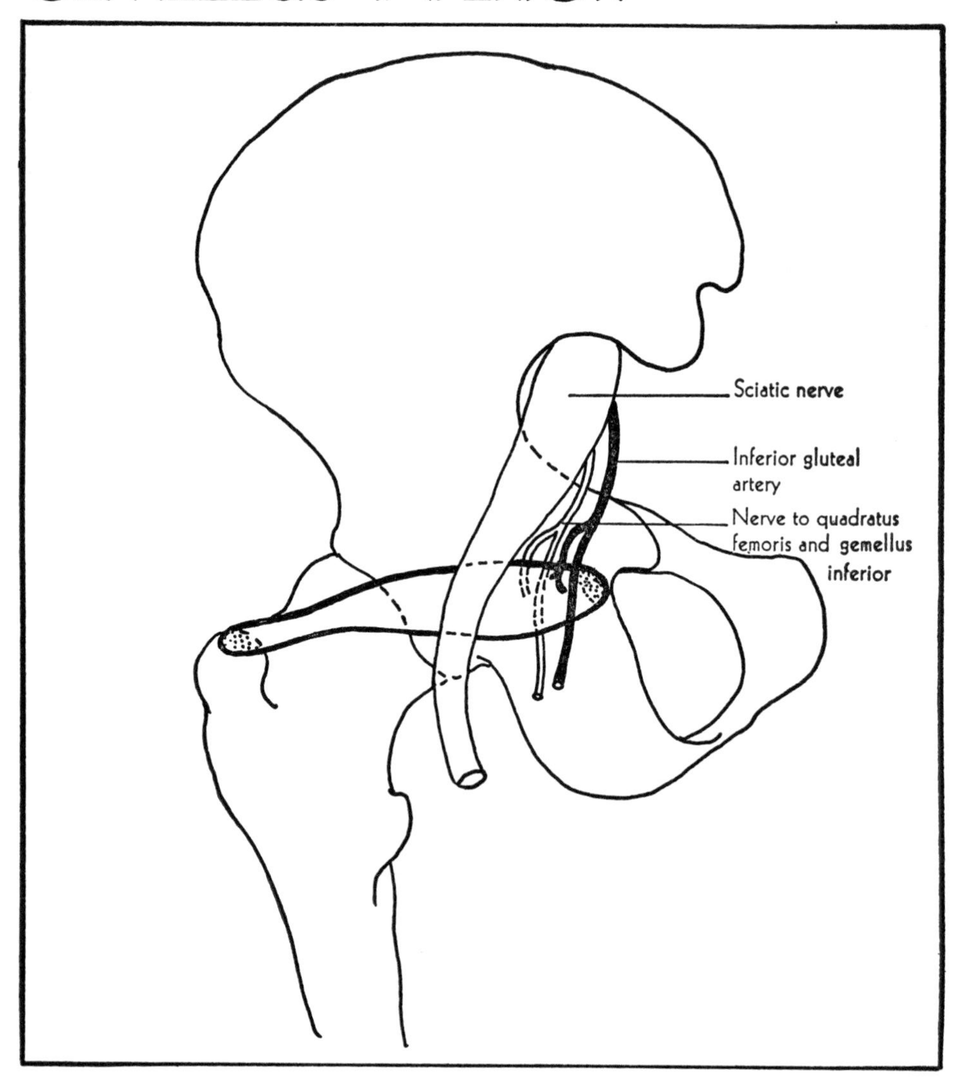

Sciatic nerve
Inferior gluteal artery
Nerve to quadratus femoris and gemellus inferior

ORIGIN: Upper part of ischial tuberosity

INSERTION: Medial surface of greater trochanter, blends with obturator internus tendon

FUNCTION: Rotates thigh laterally

NERVE: Nerve to quadratus femoris and gemellus inferior

ARTERY: Inferior gluteal

REFERENCES: GRAY		GRANT'S ATLAS
Muscle	501	272
Nerve	501, 993-994	Not shown
Artery	649	270, 271

QUADRATUS FEMORIS

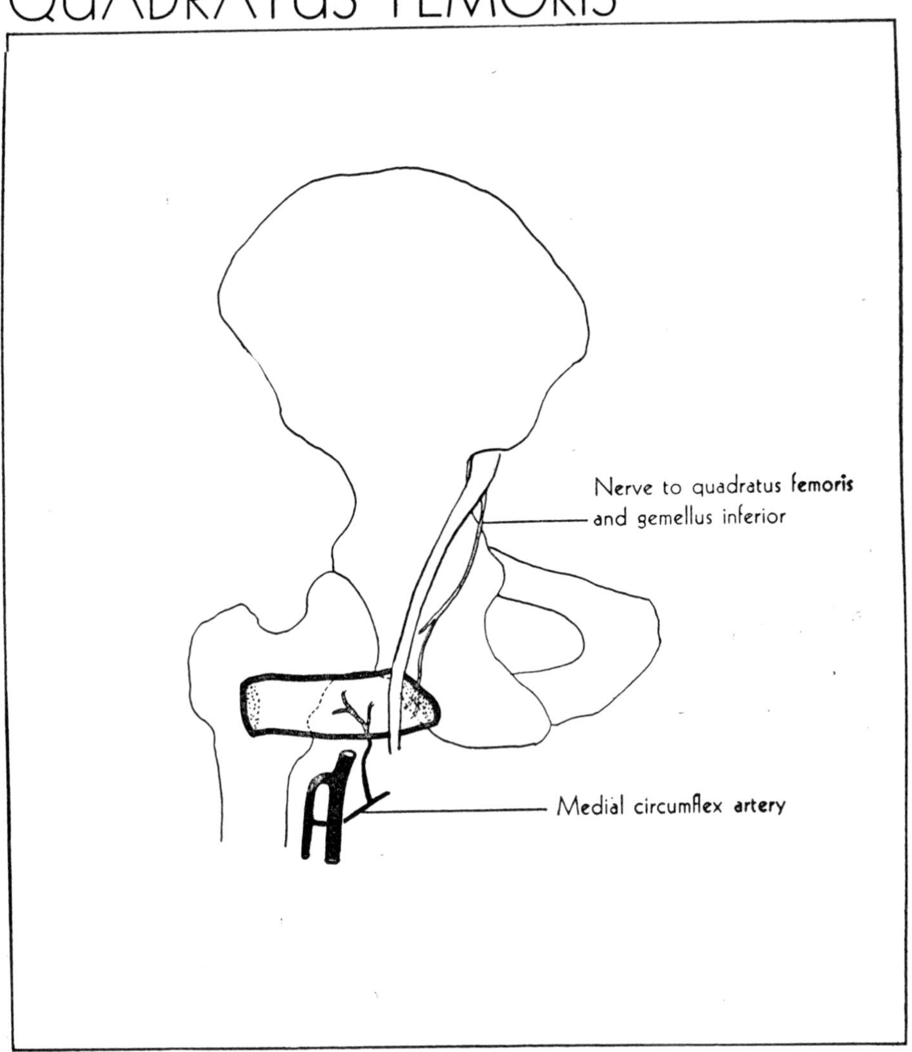
Nerve to quadratus femoris
and gemellus inferior
Medial circumflex artery

ORIGIN: Lateral margin of ischial tuberosity
INSERTION: Quadrate tubercle of femur, linea quadrata
FUNCTION: Adducts and laterally rotates thigh
NERVE: Nerve to quadratus femoris and gemellus inferior
ARTERY: Medial femoral circumflex

REFERENCES: GRAY	GRANT'S ATLAS
Muscle 501	273
Nerve 501, 993-994	Not shown
Artery 658	270, 271

OBTURATOR EXTERNUS

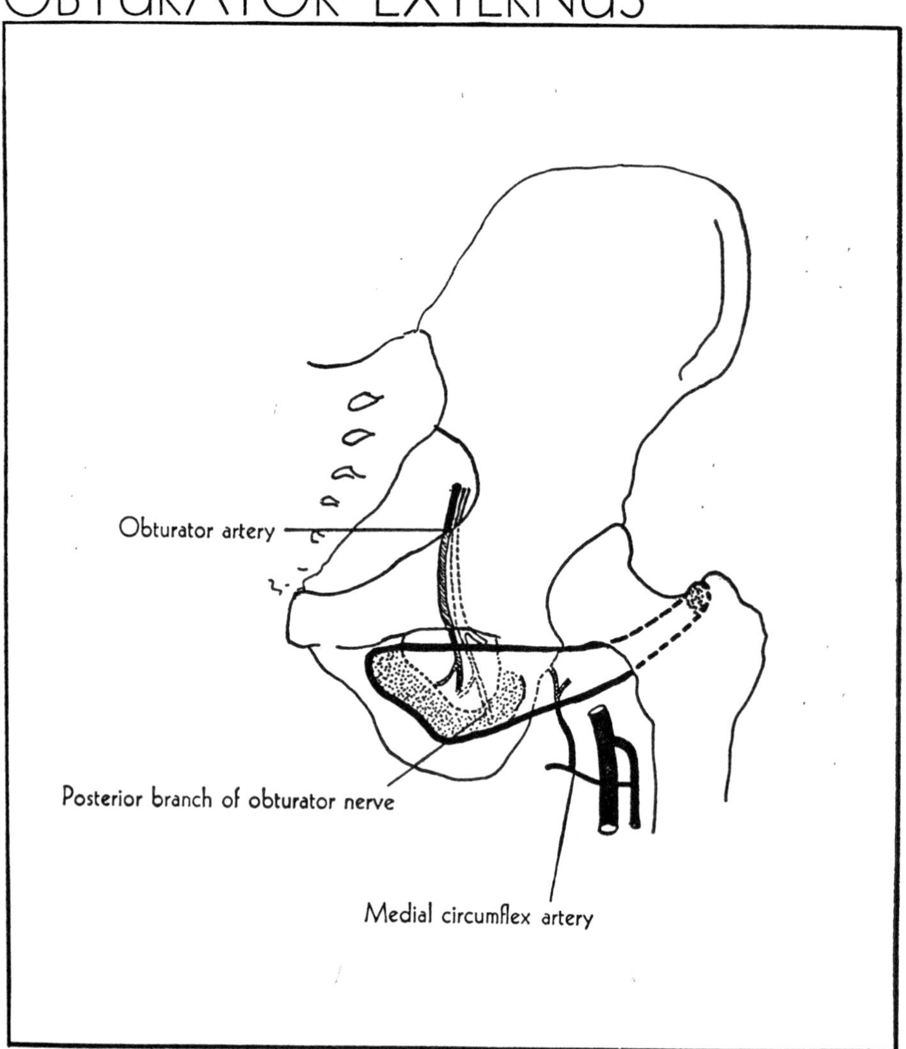

Obturator artery
Posterior branch of obturator nerve
Medial circumflex artery

ORIGIN:	Outer margin of obturator foramen, outer surface of obturator membrane
INSERTION:	Trochanteric fossa of femur
FUNCTION:	Adducts thigh, rotates it laterally
NERVE:	Posterior branch of obturator
ARTERY:	Obturator, medial femoral circumflex

REFERENCES:	GRAY	GRANT'S ATLAS
Muscle	501	274
Nerve	501, 983-985, 987	274
Artery	646, 658	271

BICEPS FEMORIS

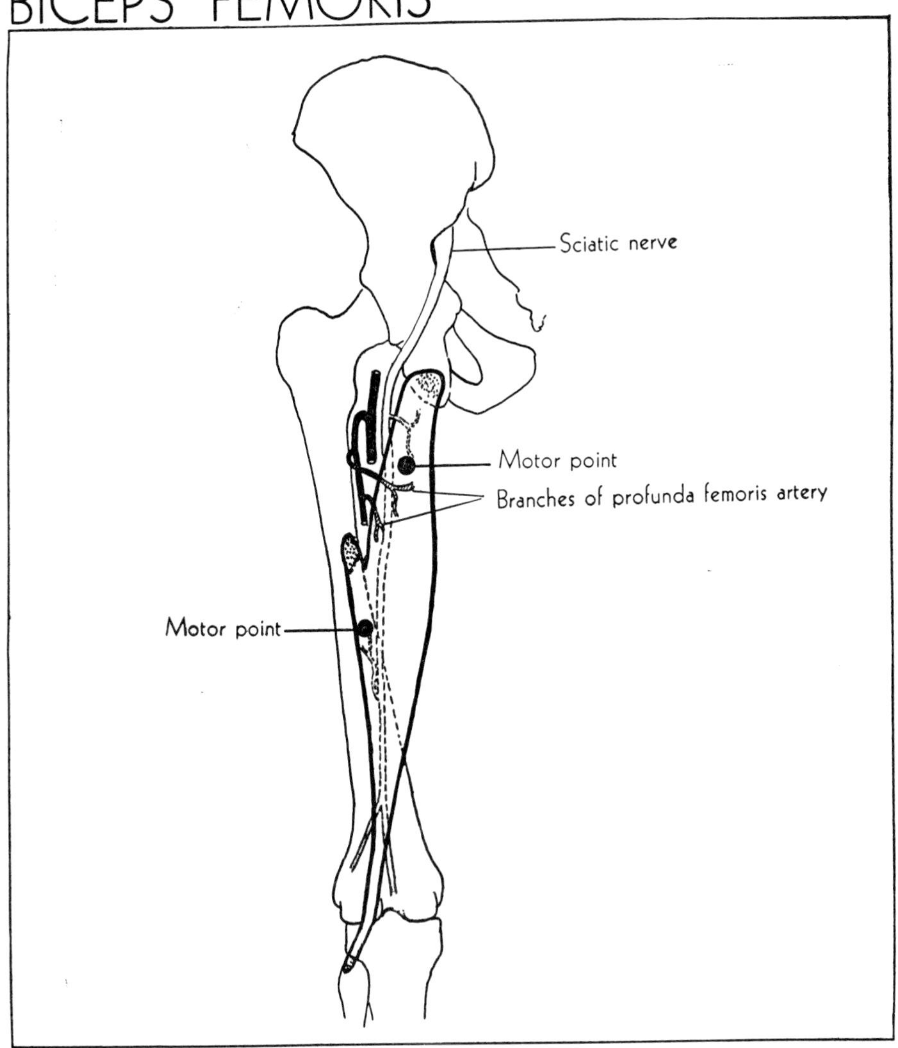

Sciatic nerve
Motor point
Branches of profunda femoris artery
Motor point

ORIGIN: Long head from ischial tuberosity and sacrotuberous ligament; short head from lateral lip of linea aspera, lateral supracondylar line of femur, lateral intermuscular septum

INSERTION: Head of fibula, lateral condyle of tibia, deep fascia on lateral side of leg

FUNCTION: Flexes leg, extends thigh, rotates leg laterally when knee is semiflexed

NERVE: Sciatic (tibial portion to long head, peroneal portion to short head)

ARTERY: Perforating branches of profunda femoris, superior muscular branches of popliteal

REFERENCES:	GRAY	GRANT'S ATLAS
Muscle	502	268
Nerve	503, 993-994, 997	270
Artery	660, 662	270, 271

SEMITENDINOSUS

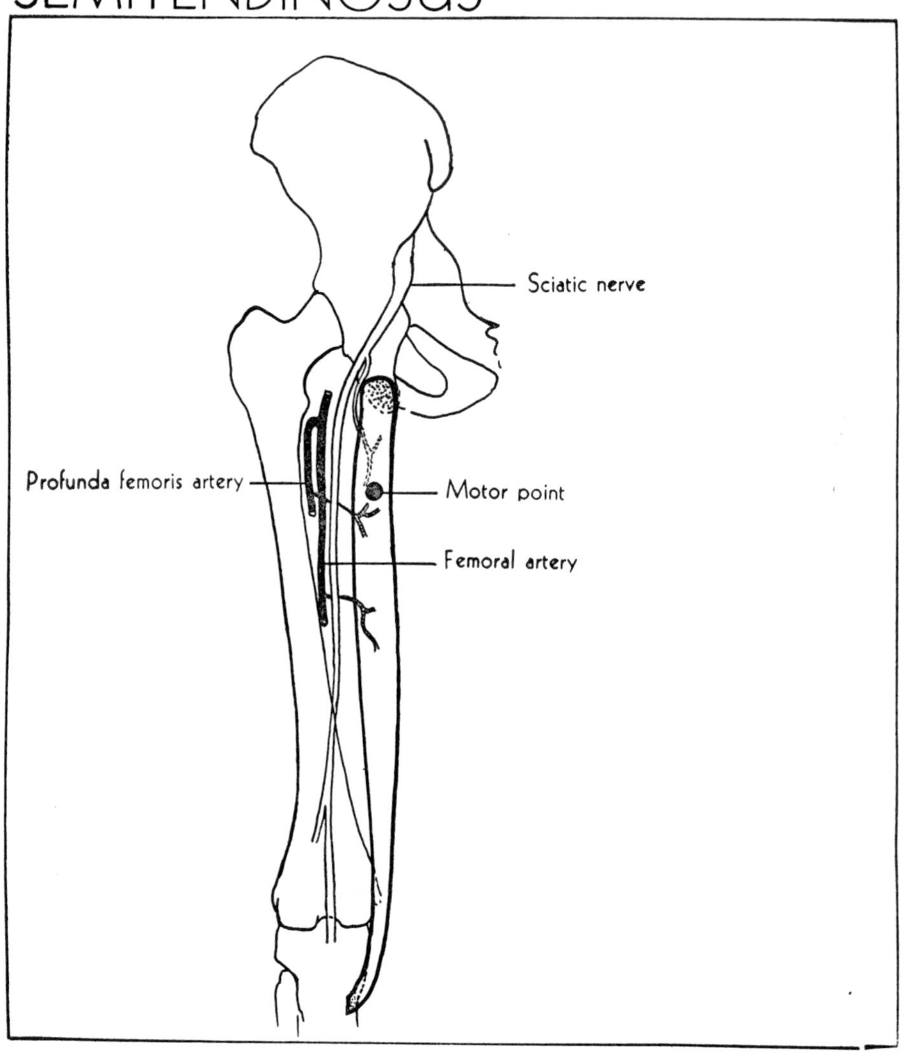

Sciatic nerve
Profunda femoris artery
Motor point
Femoral artery

ORIGIN:	Upper and medial impression of ischial tuberosity with tendon of biceps
INSERTION:	Upper part of medial surface of tibia, deep fascia of leg
FUNCTION:	Flexes leg, extends thigh, rotates leg medially when knee is semiflexed
NERVE:	Sciatic
ARTERY:	Perforating branches of profunda femoris; superior muscular branches of popliteal

REFERENCES:	GRAY	GRANT'S ATLAS
Muscle	503	268
Nerve	503, 993-994, 997	270
Artery	660, 662	Not shown

SEMIMEMBRANOSUS

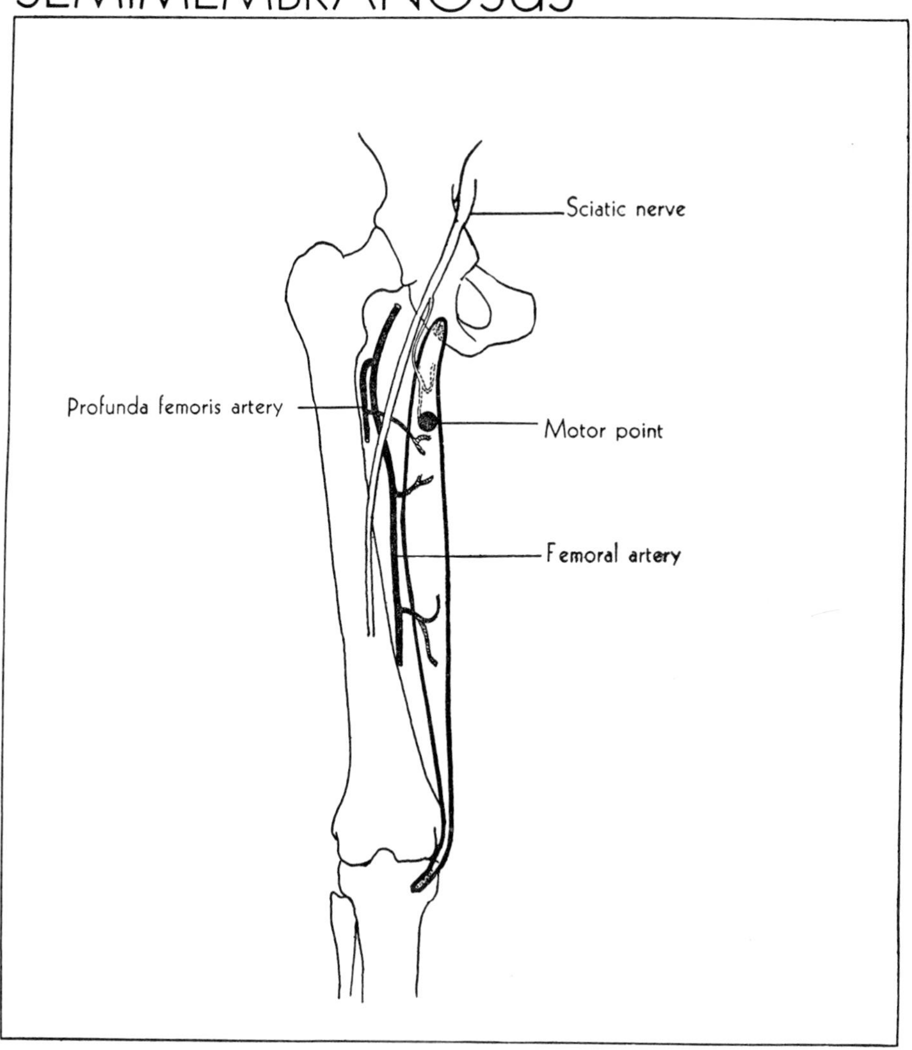

Sciatic nerve
Profunda femoris artery
Motor point
Femoral artery

ORIGIN:	Upper and lateral facet of ischial tuberosity
INSERTION:	Medial posterior surface of medial condyle of tibia
FUNCTION:	Flexes leg, extends thigh, rotates leg medially when knee is semiflexed
NERVE:	Sciatic
ARTERY:	Perforating branches of profunda femoris; superior muscular branches of popliteal

REFERENCES:	GRAY	GRANT'S ATLAS
Muscle	503	268
Nerve	503, 993-994, 997	270
Artery	660, 662	271

TIBIALIS ANTERIOR

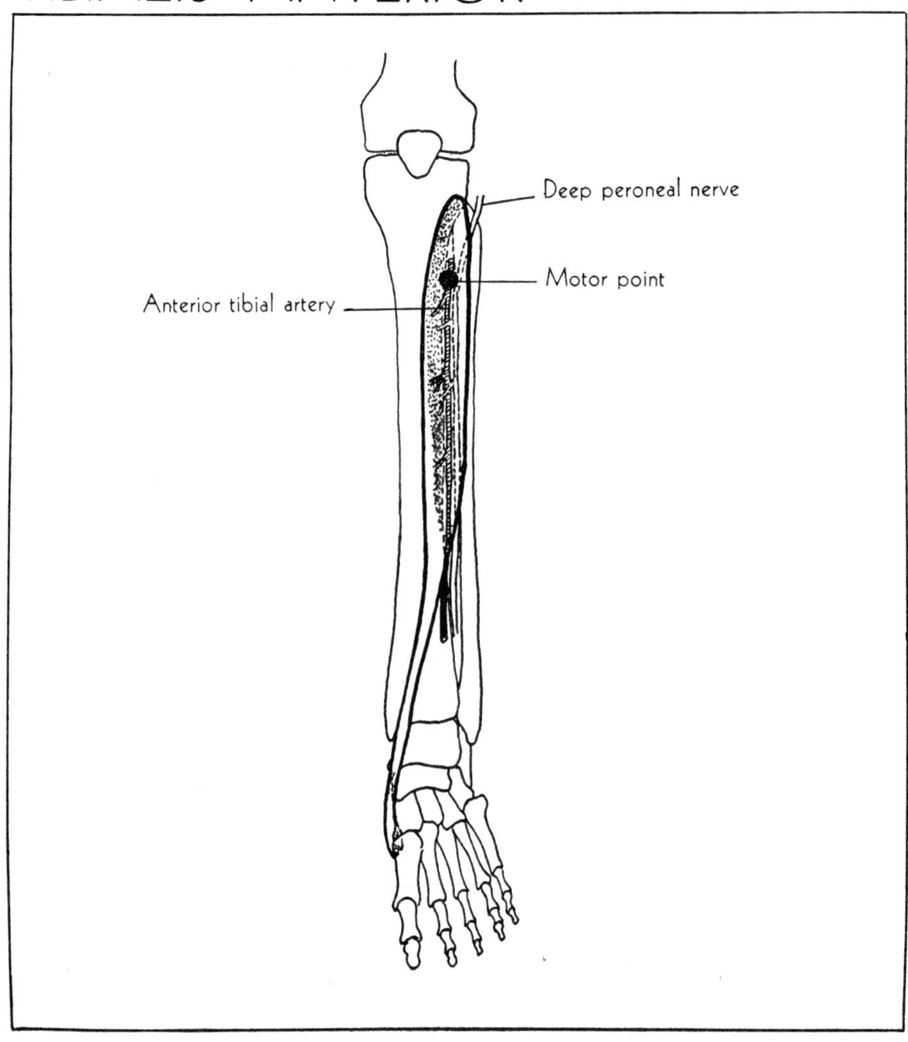
Deep peroneal nerve
Motor point
Anterior tibial artery

ORIGIN: Lateral condyle of tibia, upper two-thirds of lateral surface of tibia, interosseus membrane, deep fascia, lateral intermuscular septum

INSERTION: Medial and plantar surface of medial cuneiform bone, base of 1st metatarsal bone

FUNCTION: Dorsiflexes foot, inverts it

NERVE: Deep peroneal (anterior tibial)

ARTERY: Muscular branches of anterior tibial

REFERENCES:	GRAY	GRANT'S ATLAS
Muscle	504	305, 329
Nerve	504, 1000	305
Artery	665	305

EXTENSOR HALLUCIS LONGUS

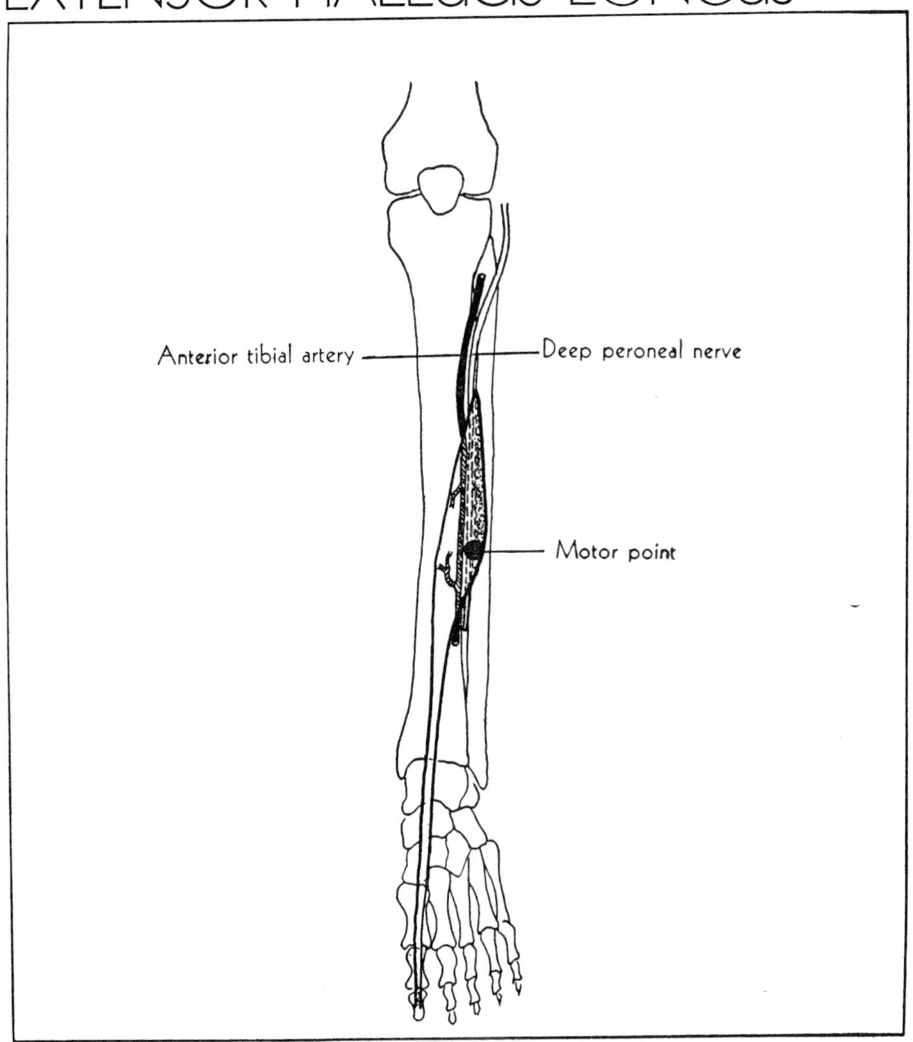
Anterior tibial artery
Deep peroneal nerve
Motor point

ORIGIN: Middle half of anterior surface of fibula, adjacent interosseous membrane
INSERTION: Base of distal phalanx of great toe
FUNCTION: Extends great toe, continued action dorsiflexes foot
NERVE: Deep peroneal (anterior tibial)
ARTERY: Muscular branches of anterior tibial

REFERENCES:	GRAY	GRANT'S ATLAS
Muscle	504	305, 307
Nerve	505, 1000	305
Artery	665	305

EXTENSOR DIGITORUM LONGUS

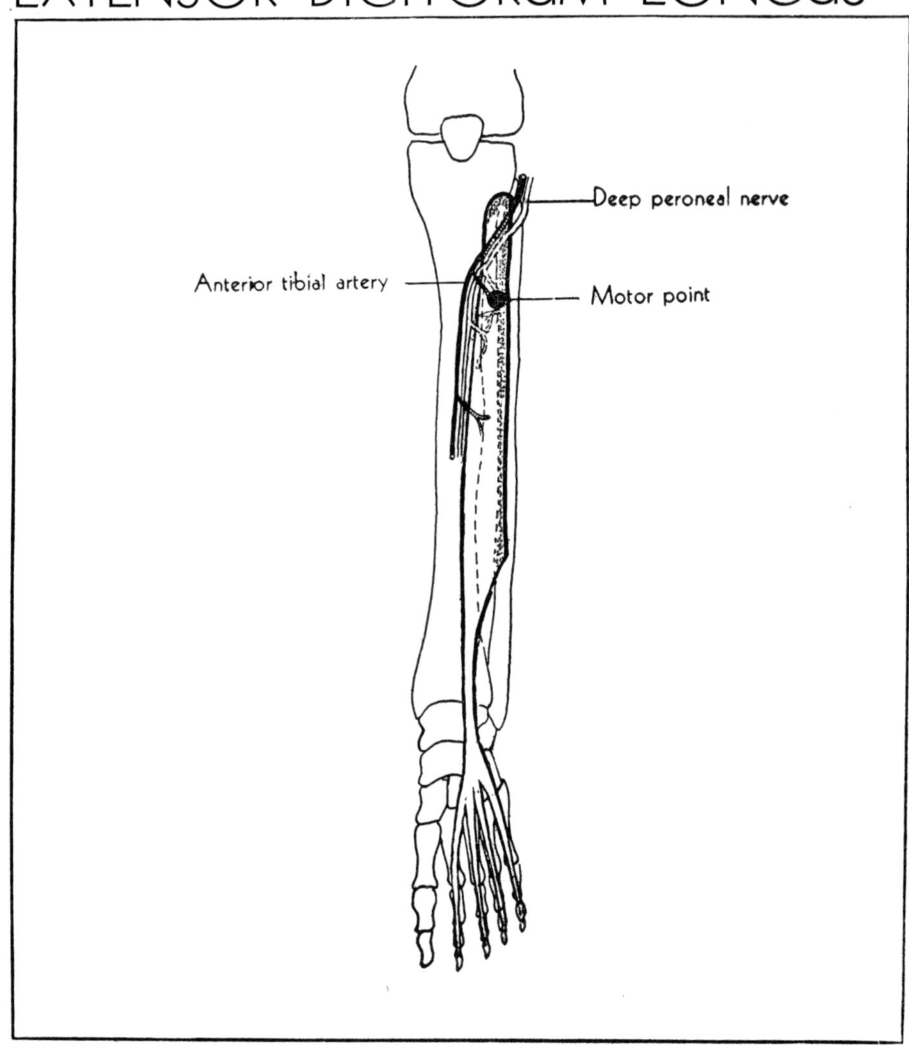
Deep peroneal nerve
Anterior tibial artery
Motor point

ORIGIN: Lateral condyle of tibia, upper three-fourths of anterior surface of fibula, interosseous membrane, deep fascia, intermuscular septa

INSERTION: Dorsal surface of middle and distal phalanges of lateral 4 toes

FUNCTION: Extends phalanges of lateral 4 toes, continued action dorsiflexes foot

NERVE: Deep peroneal (anterior tibial)

ARTERY: Muscular branches of anterior tibial

REFERENCES:	GRAY	GRANT'S ATLAS
Muscle	505	305, 307
Nerve	506, 1000	305
Artery	665	305

PERONEUS TERTIUS

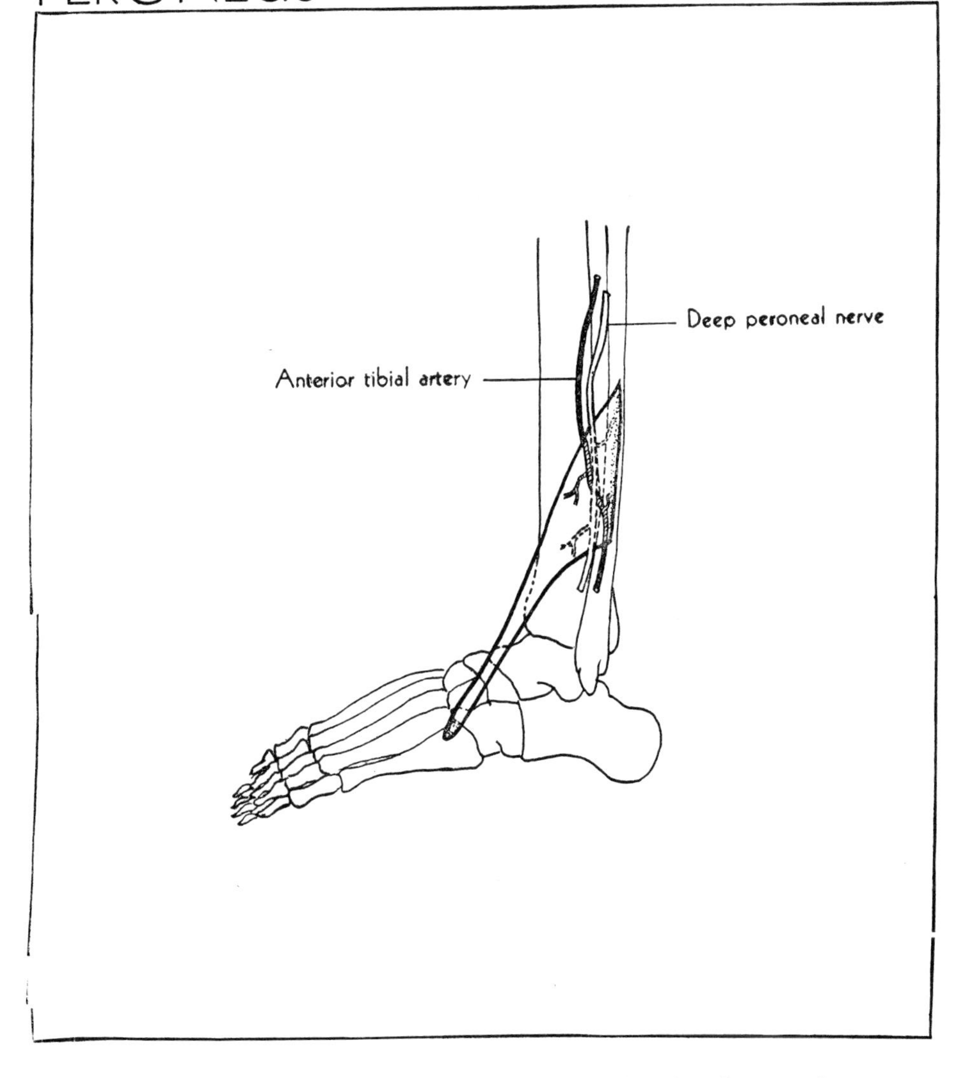
Deep peroneal nerve
Anterior tibial artery

ORIGIN:	(Peroneus tertius is commonly known as the 5th tendon of extensor digitorum longus.) Lower anterior surface of fibula, adjacent intermuscular septum
INSERTION:	Dorsal surface of base of 5th metatarsal bone
FUNCTION:	Dorsiflexes and everts foot
NERVE:	Deep peroneal (anterior tibial)
ARTERY:	Muscular branches of anterior tibial

REFERENCES:	GRAY	GRANT'S ATLAS
Muscle	506	304, 305, 308
Nerve	506, 1000	305
Artery	665	Not shown

GASTROCNEMIUS

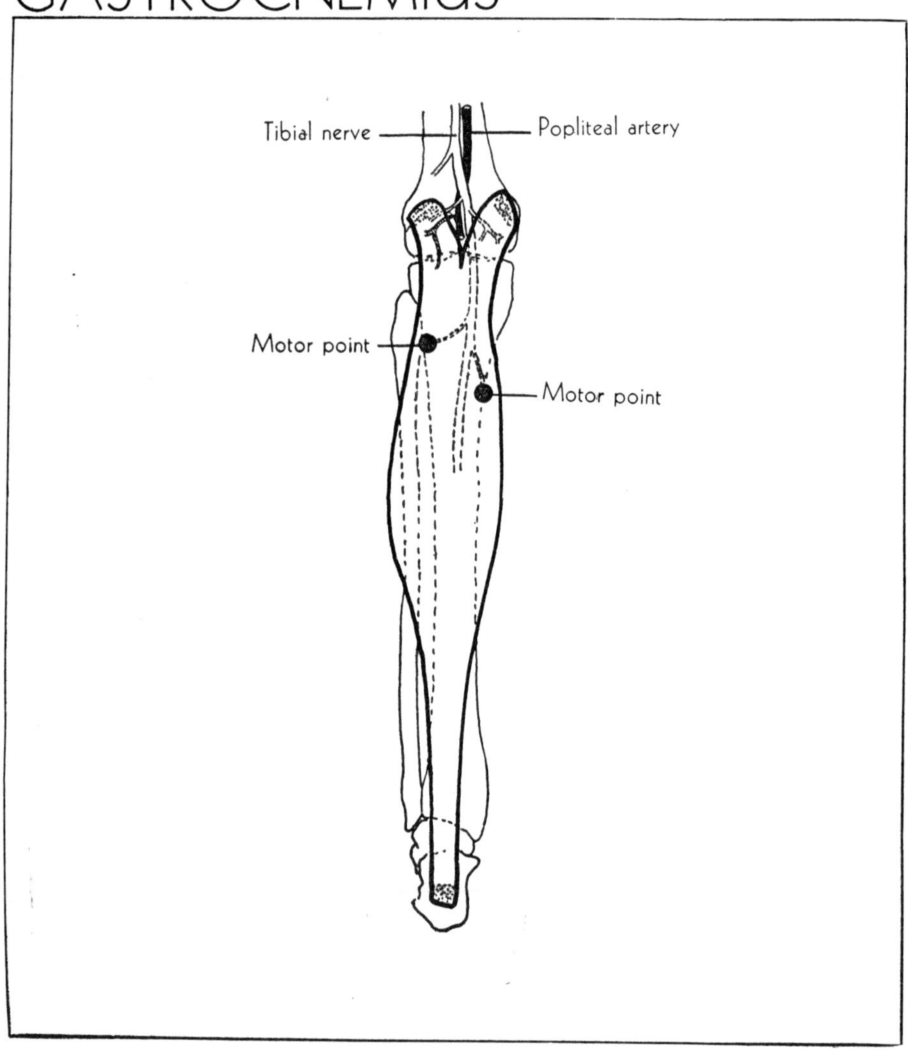
Tibial nerve
Popliteal artery
Motor point
Motor point

ORIGIN: Medial head from medial condyle and adjacent part of femur, capsule of knee joint; lateral head from lateral condyle and adjacent part of femur, capsule of knee joint

INSERTION: Into calcaneus by calcaneal tendon

FUNCTION: Plantarflexes foot; acting from below, it flexes femur on tibia

NERVE: Tibial (medial popliteal)

ARTERY: Sural branches of popliteal

REFERENCES:	GRAY	GRANT'S ATLAS
Muscle	506	317
Nerve	507, 997	286, 317
Artery	662	287

SOLEUS

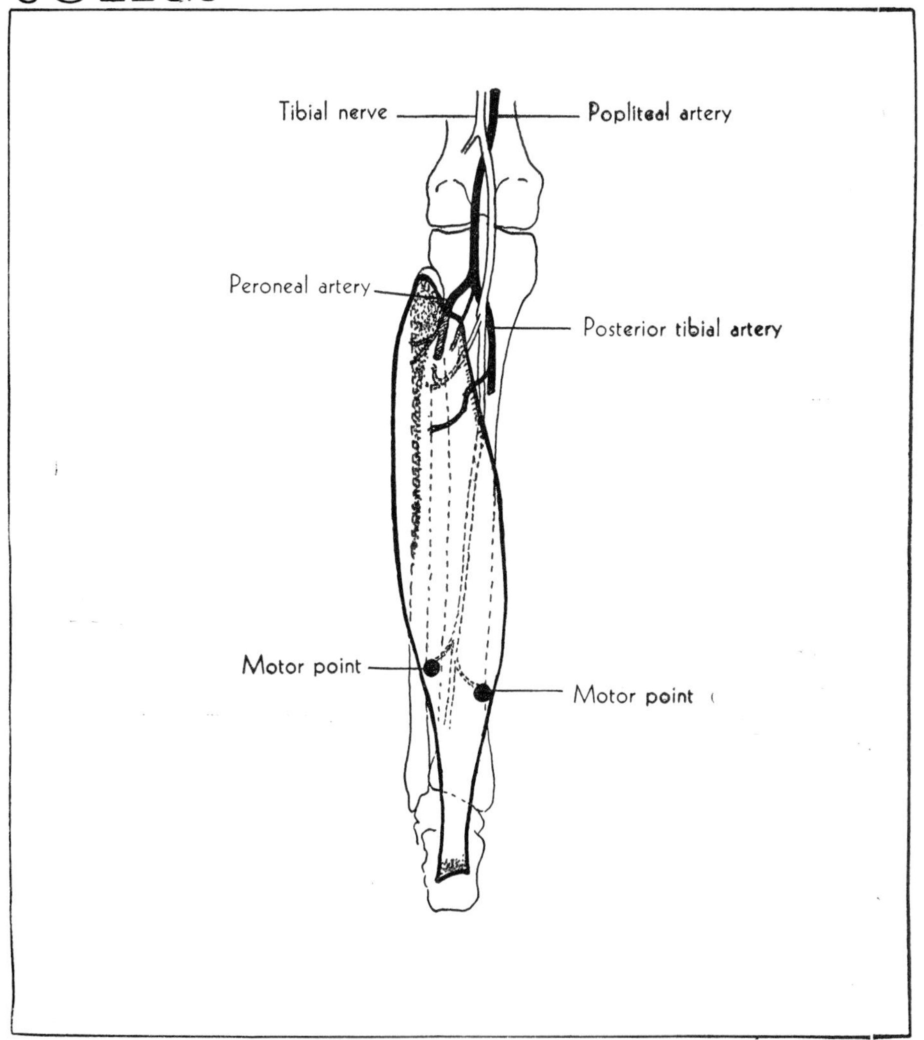
Tibial nerve
Popliteal artery
Peroneal artery
Posterior tibial artery
Motor point
Motor point

ORIGIN: Posterior surface of head and upper third of shaft of fibula, middle third of medial border of tibia, tendinous arch between tibia and fibula

INSERTION: Into calcaneus by calcaneal tendon

FUNCTION: Plantarflexes foot, steadies leg upon foot

NERVE: Tibial (medial popliteal)

ARTERY: Posterior tibial, peroneal, sural branches of popliteal

REFERENCES:	GRAY	GRANT'S ATLAS
Muscle	507	318
Nerve	507, 997	286, 318
Artery	662, 667-668	318

PLANTARIS

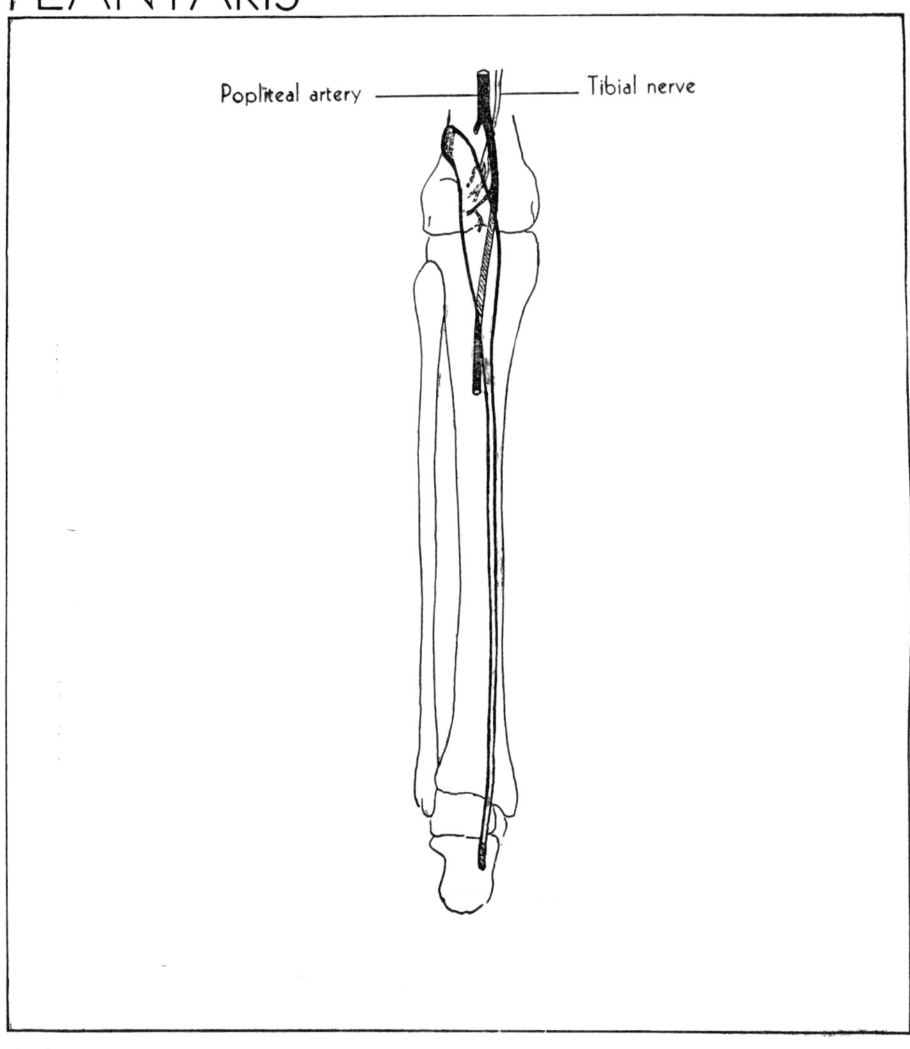

Popliteal artery
Tibial nerve

ORIGIN: Lateral supracondylar line of femur, oblique popliteal ligament of knee joint
INSERTION: Medial side of posterior part of calcaneus, calcaneal tendon
FUNCTION: Plantarflexes foot
NERVE: Tibial (medial popliteal)
ARTERY: Sural branches of popliteal

REFERENCES:	GRAY	GRANT'S ATLAS
Muscle	507	286
Nerve	507, 997	Not shown
Artery	662	Not shown

POPLITEUS

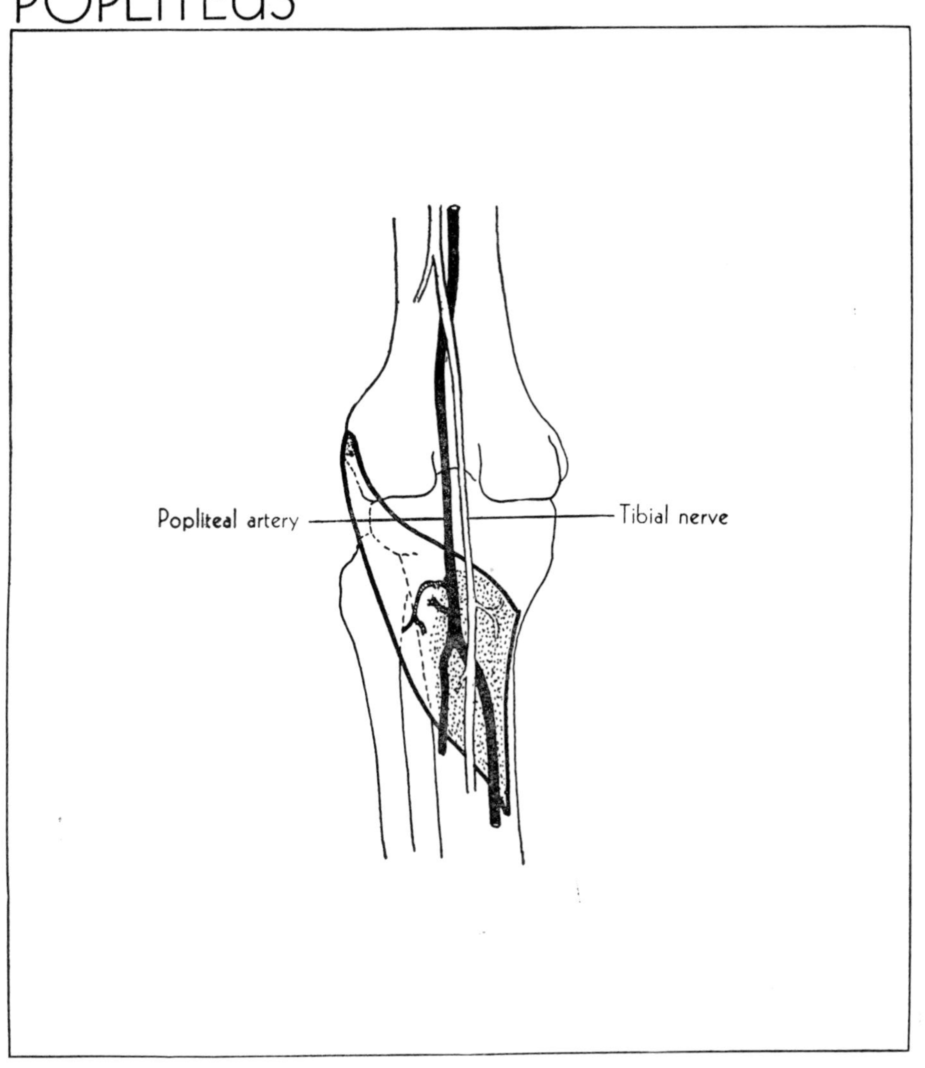
Popliteal artery
Tibial nerve

ORIGIN: Lateral condyle of femur, oblique popliteal ligament of knee
INSERTION: Triangular area on posterior surface of tibia above soleal line
FUNCTION: Flexes leg, rotates tibia medially at beginning of flexion
NERVE: Tibial (medial or internal popliteal)
ARTERY: Genicular branches of popliteal

REFERENCES: GRAY		GRANT'S ATLAS
Muscle	508	287, 302, 320
Nerve	508, 997	286, 320
Artery	664	287

FLEXOR HALLUCIS LONGUS

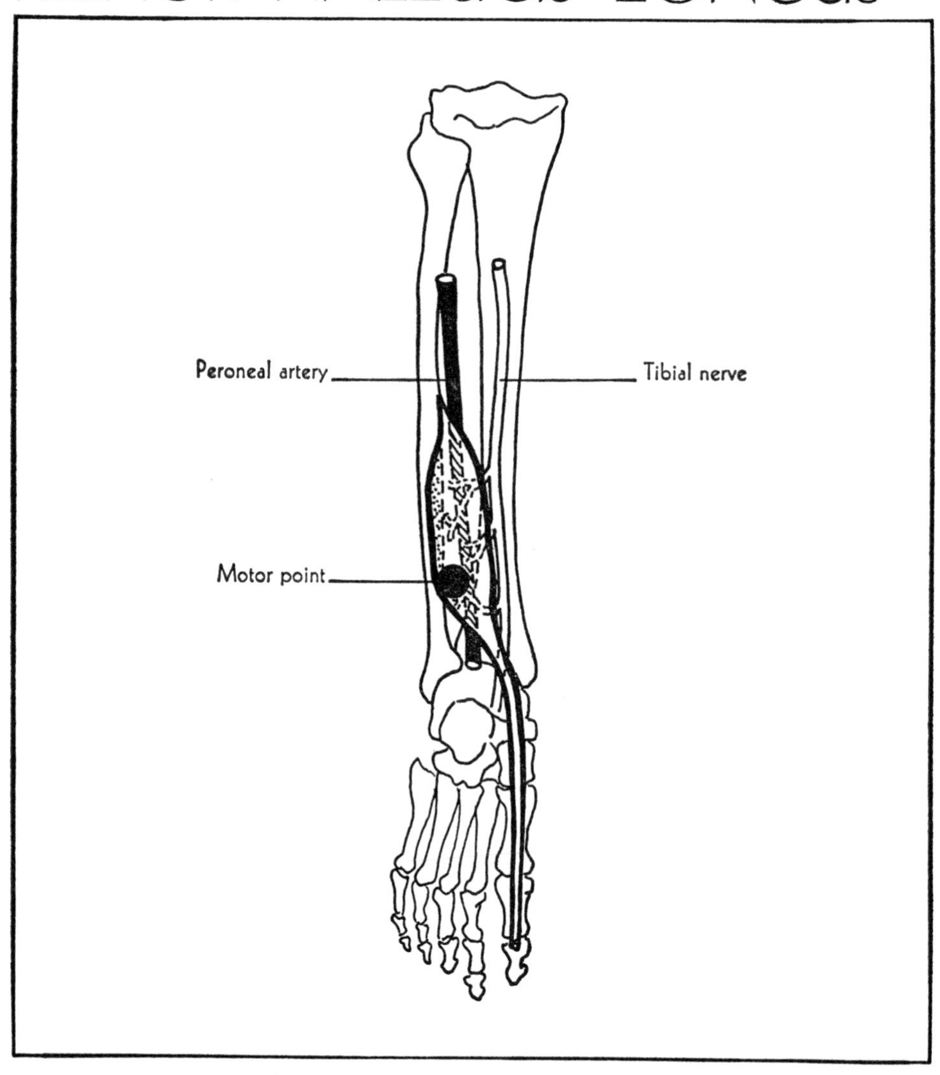

Peroneal artery
Tibial nerve
Motor point

ORIGIN:	Lower two-thirds of posterior surface of fibula, interosseous membrane, adjacent intermuscular septa and fascia
INSERTION:	Base of distal phalanx of great toe
FUNCTION:	Flexes great toe, continued action aids in plantarflexing foot
NERVE:	Tibial (medial or internal popliteal)
ARTERY:	Muscular branches of peroneal

REFERENCES: GRAY		GRANT'S ATLAS
Muscle	508	319, 322
Nerve	509, 997-998	320
Artery	668	319

FLEXOR DIGITORUM LONGUS

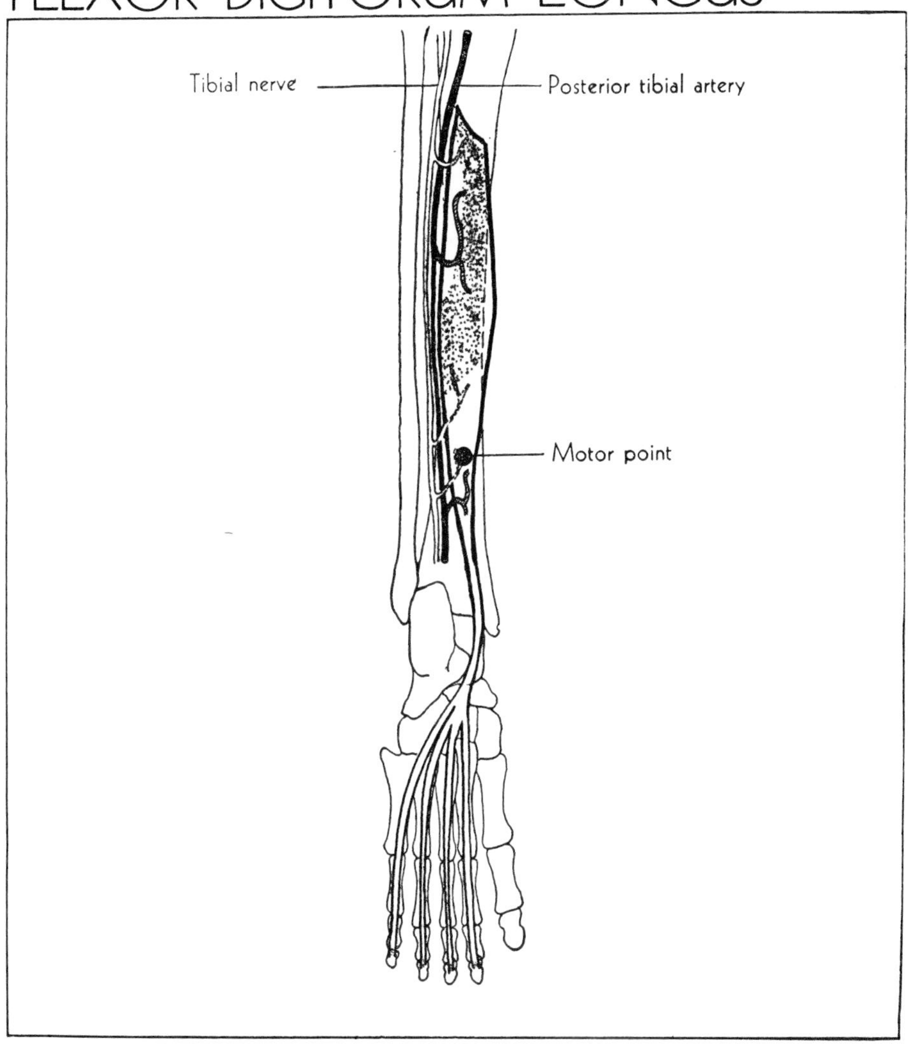
Tibial nerve
Posterior tibial artery
Motor point

ORIGIN: Posterior surface of middle three-fifths of tibia, fascia covering tibialis posterior

INSERTION: Plantar surface of base of distal phalanx of lateral 4 toes

FUNCTION: Flexes phalanges of lateral 4 toes, continued action plantarflexes foot

NERVE: Tibial (medial or internal popliteal)

ARTERY: Posterior tibial

REFERENCES:	GRAY	GRANT'S ATLAS
Muscle	509	320, 328
Nerve	509, 997-998	320
Artery	668	319

TIBIALIS POSTERIOR

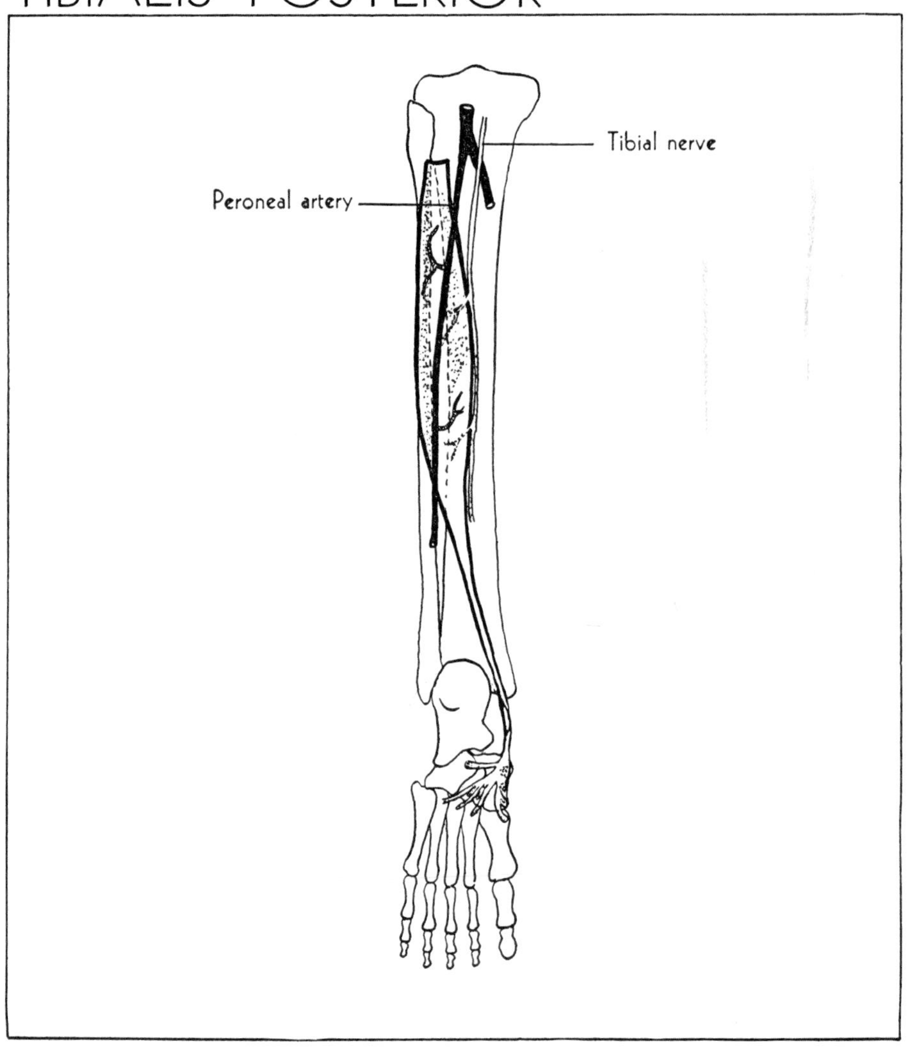

Tibial nerve
Peroneal artery

ORIGIN: Lateral part of posterior surface of tibia, upper two-thirds of medial surface of fibula, deep transverse fascia, adjacent intermuscular septa, posterior surface of interosseus membrane

INSERTION: Tuberosity of navicular bone, plantar surface of all cuneiform bones, plantar surface of base of 2d, 3d, and 4th metatarsal bones, cuboid bone, sustentaculum tali

FUNCTION: Plantarflexes foot, inverts it

NERVE: Tibial (medial or internal popliteal)

ARTERY: Peroneal

REFERENCES:	GRAY	GRANT'S ATLAS
Muscle	509	320, 329
Nerve	509, 997-998	320
Artery	668	320

PERONEUS LONGUS

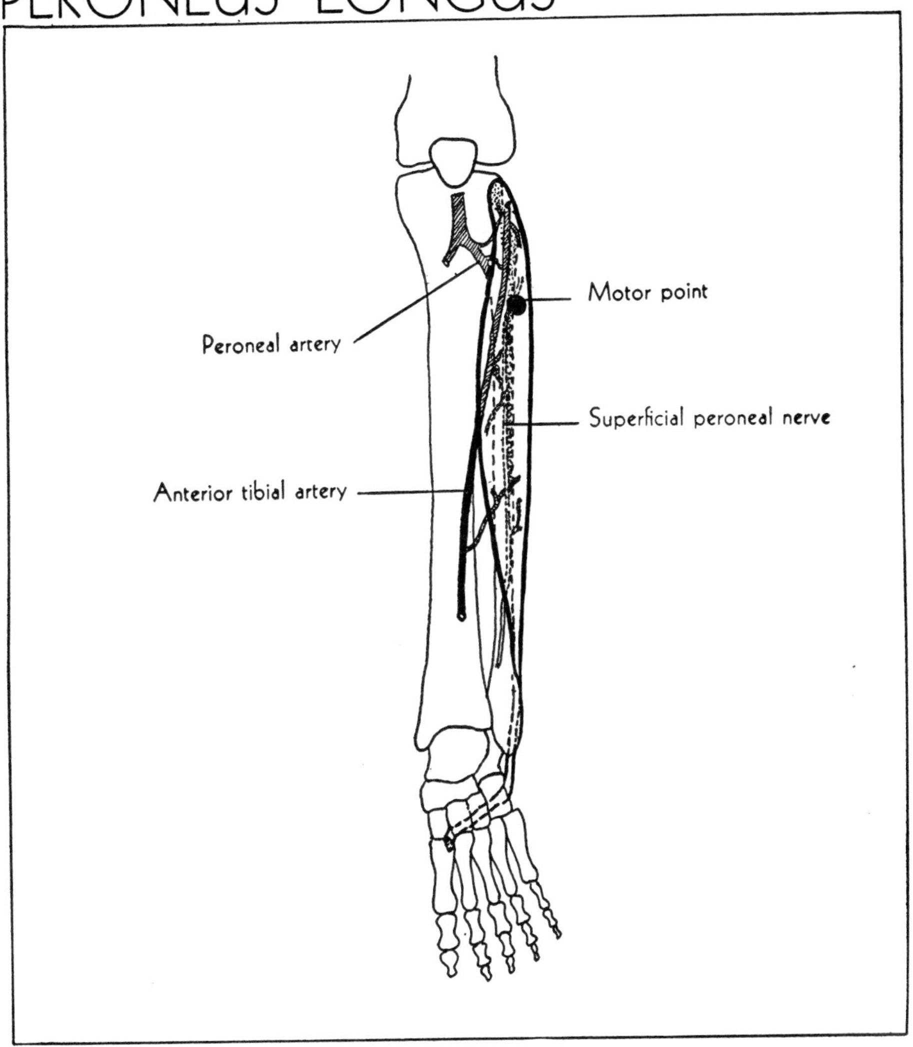

Motor point
Peroneal artery
Superficial peroneal nerve
Anterior tibial artery

ORIGIN: Lateral condyle of tibia, head and upper two-thirds of lateral surface of fibula, adjacent fascia, intermuscular septa

INSERTION: Lateral side of medial cuneiform bone, base of 1st metatarsal bone

FUNCTION: Plantarflexes foot, everts it

NERVE: Superficial peroneal (musculocutaneous)

ARTERY: Muscular branches of anterior tibial, muscular branches of peroneal

REFERENCES: GRAY		GRANT'S ATLAS
Muscle	510	304, 308, 341
Nerve	511, 1000	306
Artery	665, 668	Not shown

PERONEUS BREVIS

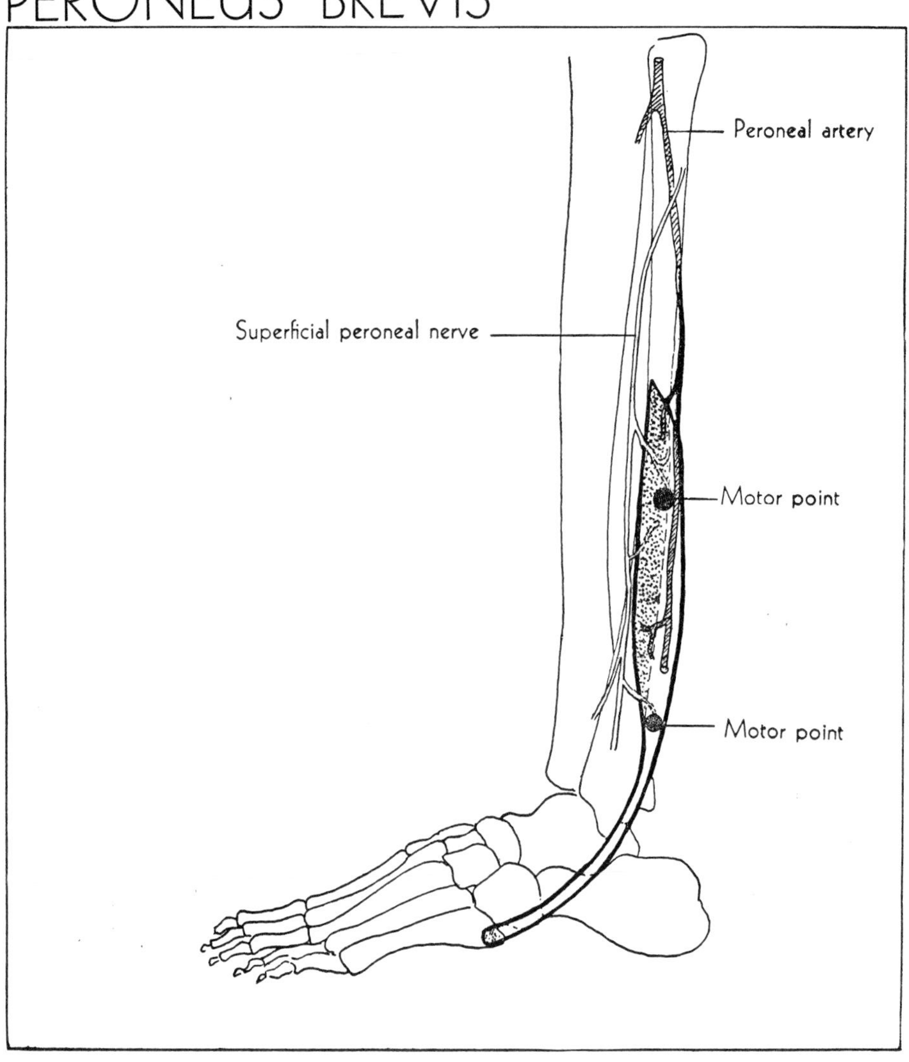

Peroneal artery
Superficial peroneal nerve
Motor point
Motor point

ORIGIN:	Lower two-thirds of lateral surface of fibula, adjacent intermuscular septa
INSERTION:	Lateral side of base of 5th metatarsal bone
FUNCTION:	Plantarflexes foot, everts it
NERVE:	Superficial peroneal (musculocutaneous)
ARTERY:	Muscular branches of peroneal

REFERENCES:	GRAY	GRANT'S ATLAS
Muscle	511	304, 308
Nerve	511, 1000	306
Artery	668	Not shown

EXTENSOR DIGITORUM BREVIS

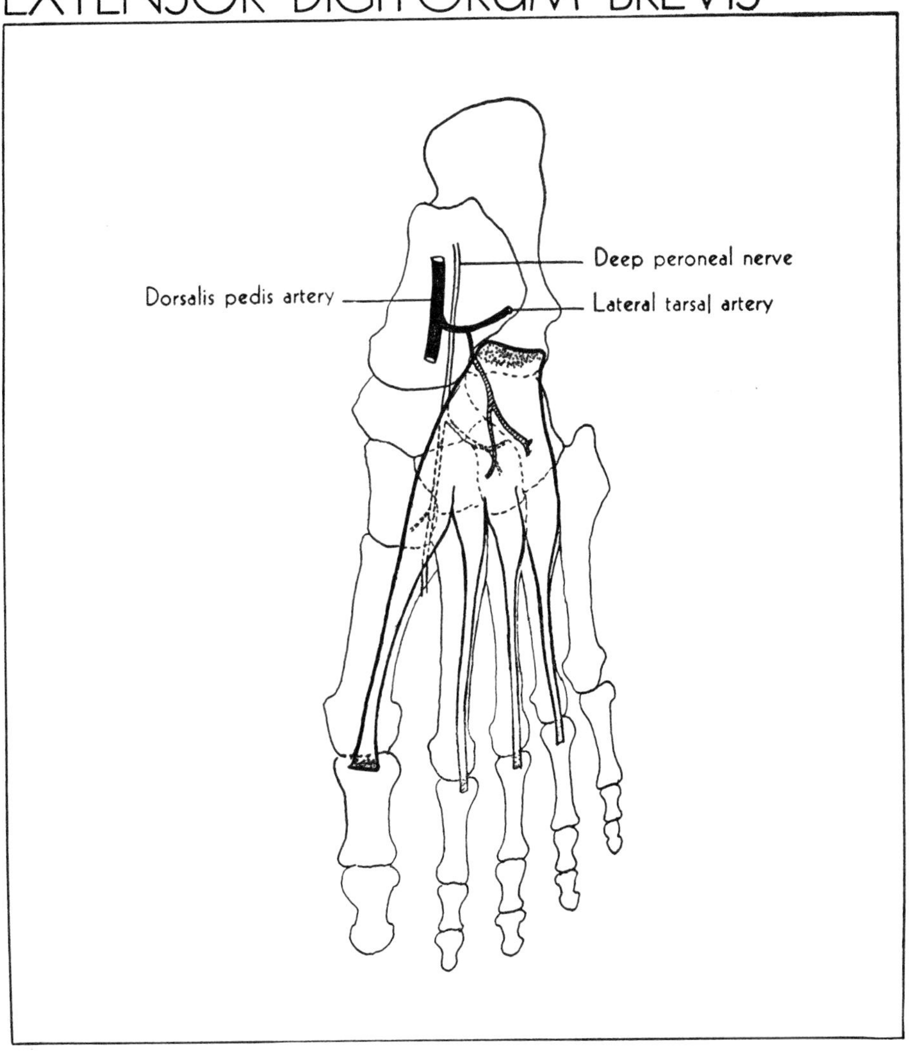
Deep peroneal nerve
Dorsalis pedis artery
Lateral tarsal artery

ORIGIN: Forepart of upper and lateral surface of calcaneus, lateral talocalcaneal ligament, cruciate crural ligament

INSERTION: 1st tendon into dorsal surface of base of proximal phalanx of great toe, remaining 3 tendons into lateral sides of tendons of extensor digitorum longus

FUNCTION: Extends phalanges of 4 medial toes

NERVE: Deep peroneal (anterior tibial)

ARTERY: Dorsalis pedis, lateral tarsal

REFERENCES:	GRAY	GRANT'S ATLAS
Muscle	514	307, 309
Nerve	514, 1000	305
Artery	666	Not shown

ABDUCTOR HALLUCIS

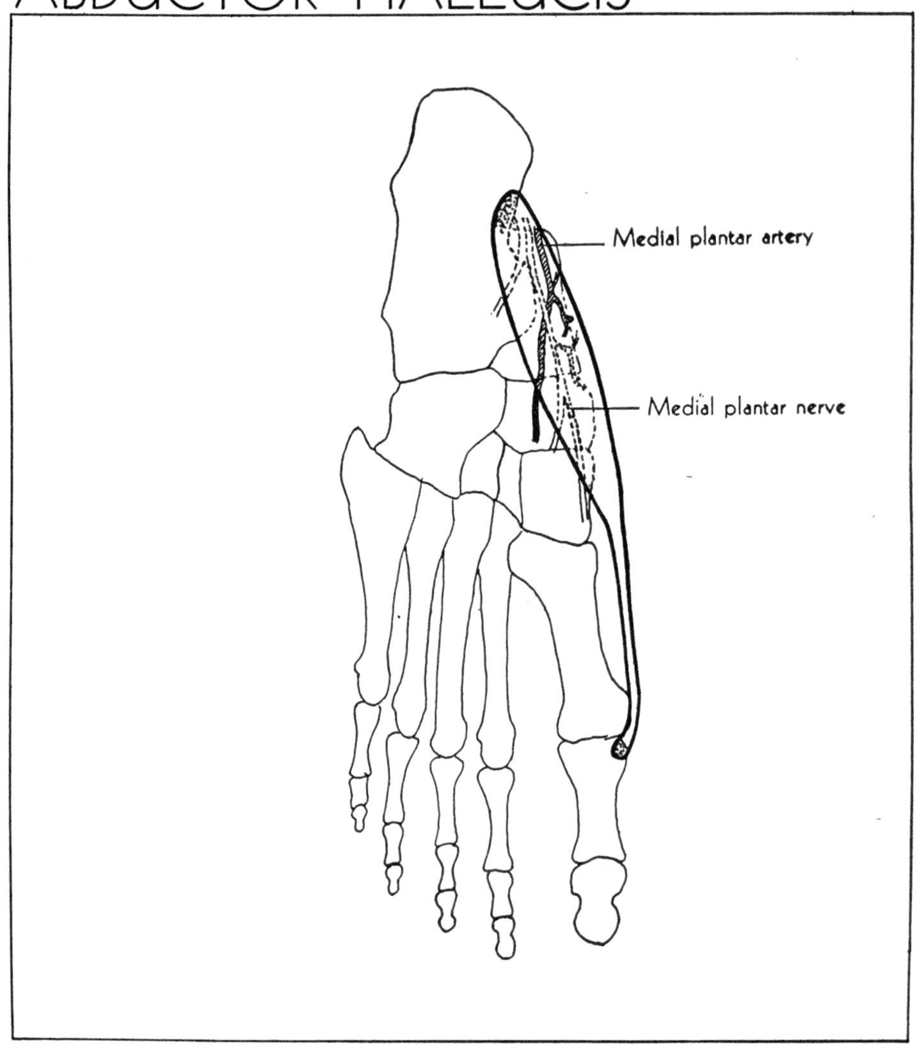
Medial plantar artery
Medial plantar nerve

ORIGIN: Medial process of calcaneus, laciniate ligament, plantar aponeurosis, adjacent intermuscular septum

INSERTION: Medial side of base of proximal phalanx of great toe

FUNCTION: Abducts great toe

NERVE: Medial plantar

ARTERY: Medial plantar

REFERENCES:	GRAY	GRANT'S ATLAS
Muscle	515	331
Nerve	515, 999	321, 330
Artery	668	321

FLEXOR DIGITORUM BREVIS

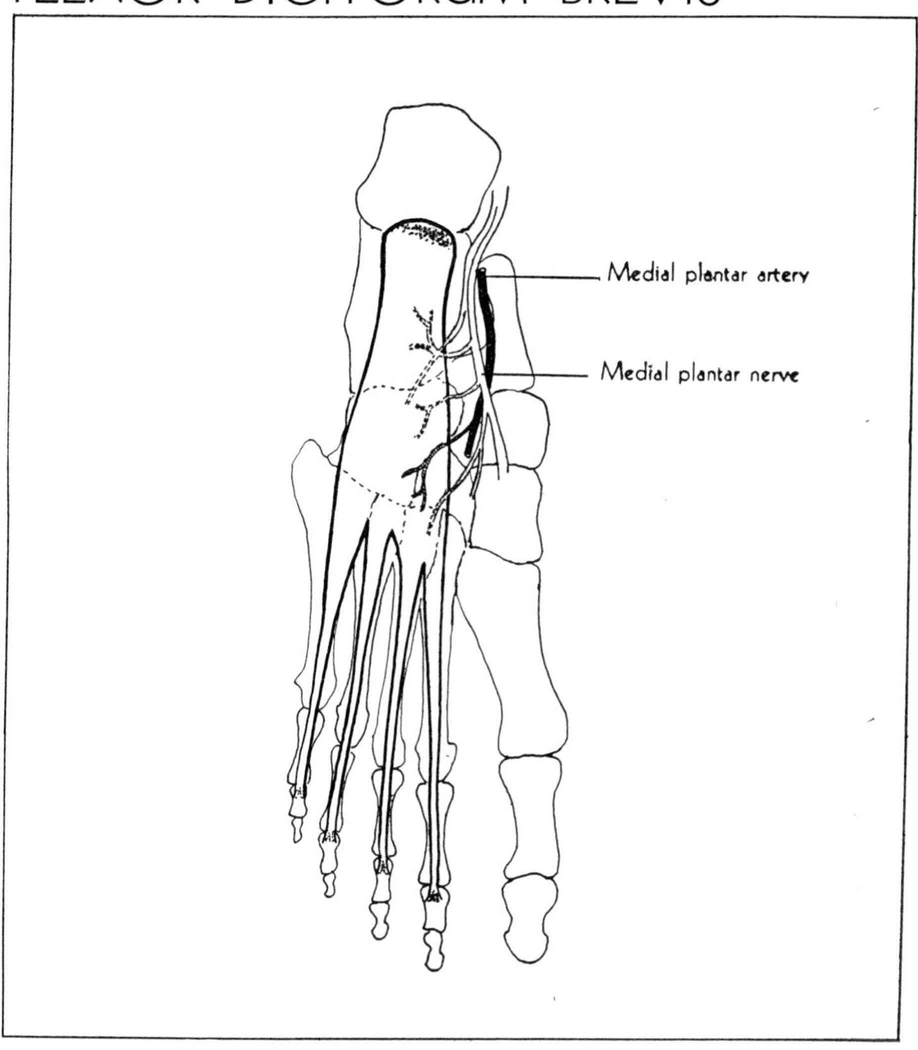

Medial plantar artery
Medial plantar nerve

ORIGIN: Medial process of calcaneus, plantar aponeurosis, adjacent intermuscular septa

INSERTION: Middle phalanx of lateral 4 toes

FUNCTION: Flexes middle phalanges on proximal, continued action also flexes proximal phalanges of lateral 4 toes

NERVE: Medial plantar

ARTERY: Medial plantar

REFERENCES:	GRAY	GRANT'S ATLAS
Muscle	516	327
Nerve	516, 999	Not shown
Artery	668	Not shown

ABDUCTOR DIGITI MINIMI

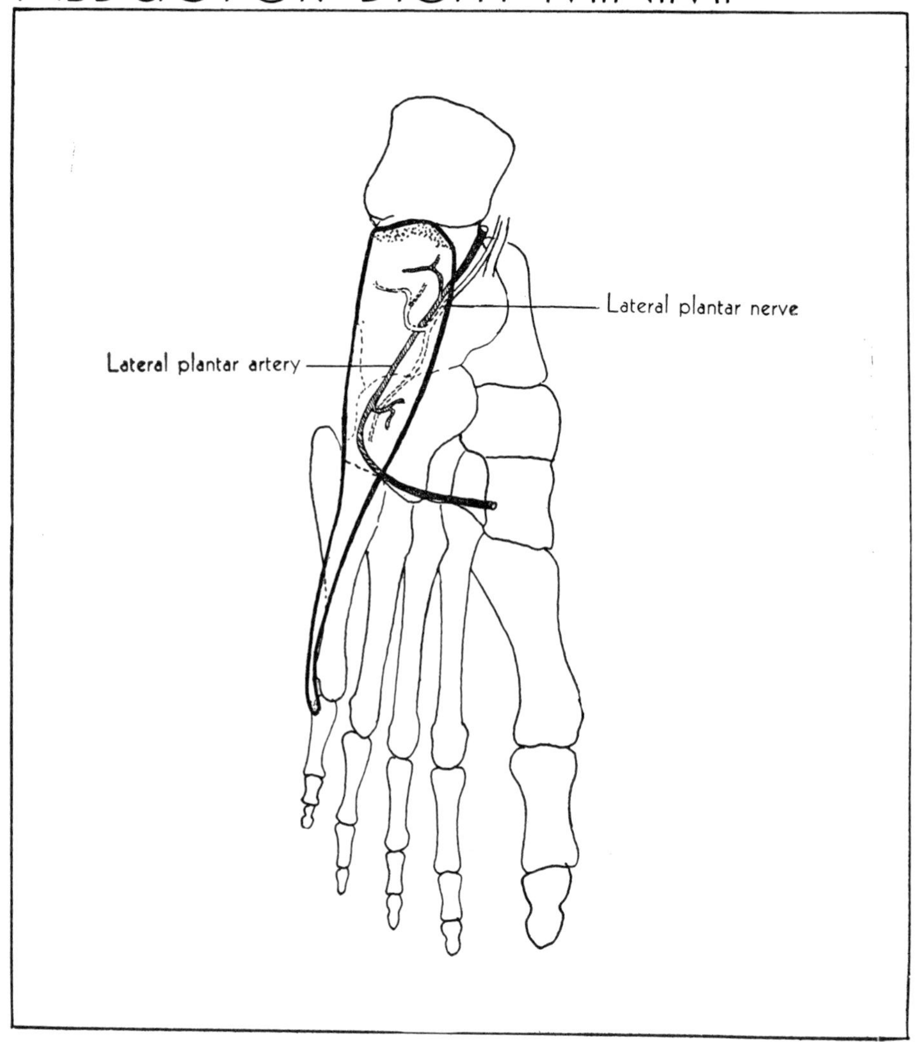
Lateral plantar nerve
Lateral plantar artery

ORIGIN: Lateral and medial processes of calcaneus, calcaneal fascia, adjacent intermuscular septum

INSERTION: Lateral side of base of proximal phalanx of little toe

FUNCTION: Abducts little toe, assists in flexing it

NERVE: Lateral plantar

ARTERY: Lateral plantar

REFERENCES:	GRAY	GRANT'S ATLAS
Muscle	516	331
Nerve	516, 999	Not shown
Artery	668	Not shown

QUADRATUS PLANTAE (Flexor Accessorius)

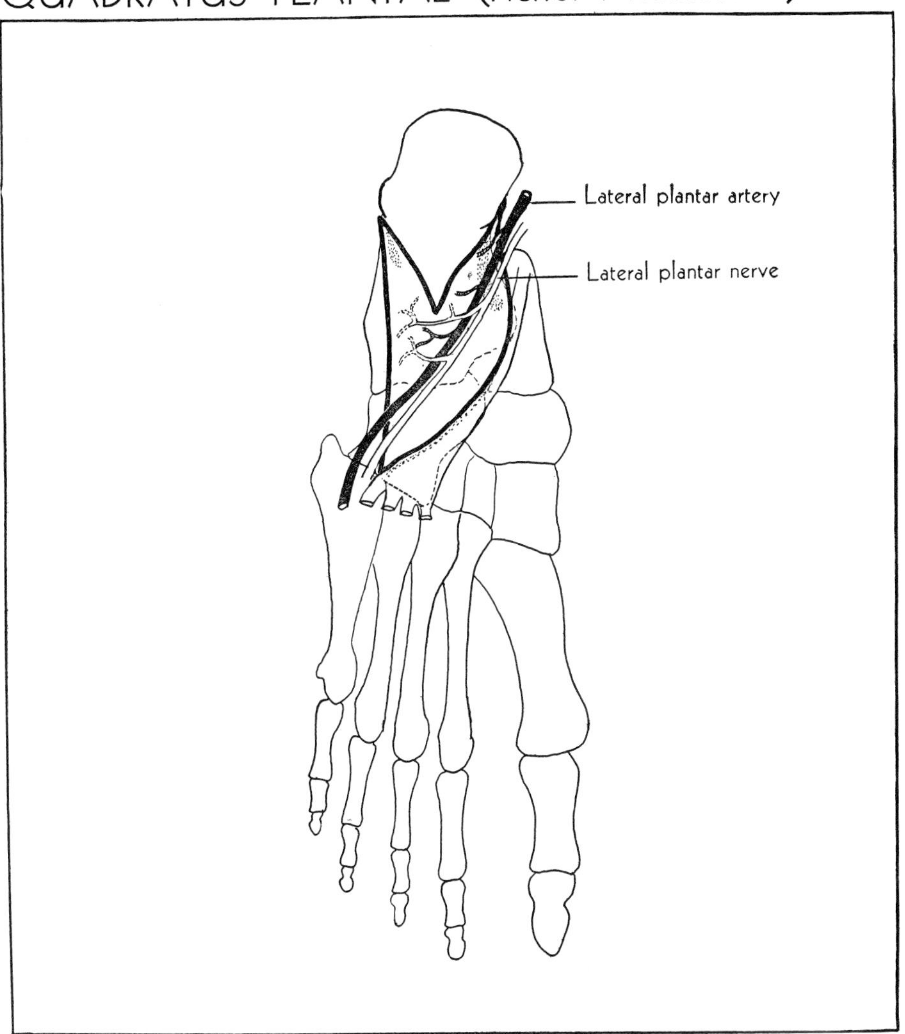

Lateral plantar artery
Lateral plantar nerve

ORIGIN: Medial head from medial surface of calcaneus and medial border of long plantar ligament, lateral head from lateral border of plantar surface of calcaneus and lateral border of long plantar ligament

INSERTION: Tendons of flexor digitorum longus

FUNCTION: Flexes terminal phalanges of lateral 4 toes

NERVE: **Lateral plantar**

ARTERY: **Lateral plantar**

REFERENCES: GRAY		GRANT'S ATLAS
Muscle	516	328, 329
Nerve	517, 999	330
Artery	668	330

LUMBRICALES

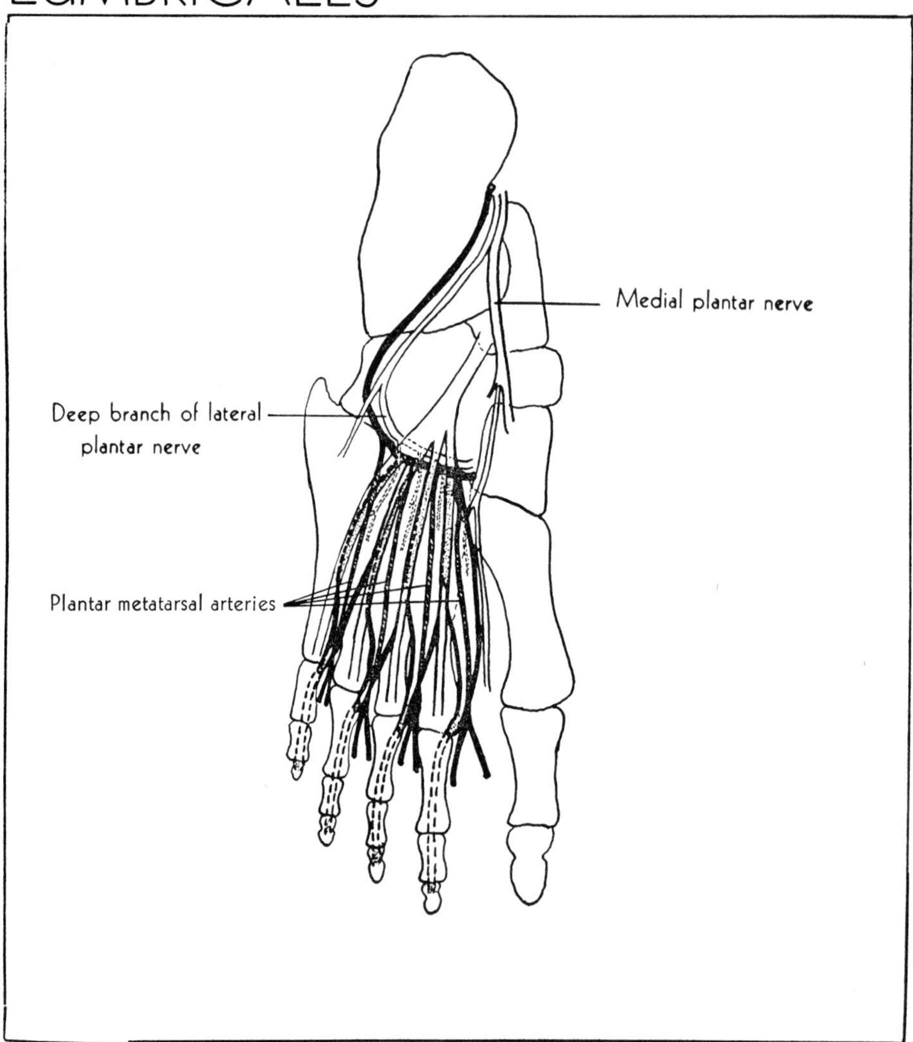

Medial plantar nerve
Deep branch of lateral plantar nerve
Plantar metatarsal arteries

ORIGIN: There are 4 lumbricales, all arising from tendons of flexor digitorum longus: 1st from medial side of tendon for 2d toe, 2d from adjacent sides of tendons for 2d and 3d toes, 3d from adjacent sides of tendons for 3d and 4th toes, 4th from adjacent sides of tendons for 4th and 5th toes

INSERTION: With tendons of extensor digitorum longus and interossei into bases of terminal phalanges of 4 lateral toes

FUNCTION: Flex toes at metatarsophalangeal joints, extend toes at interphalangeal joints

NERVE: Medial plantar, deep lateral plantar

ARTERY: Plantar metatarsal

REFERENCES:	GRAY	GRANT'S ATLAS
Muscle	517	328
Nerve	517, 999	Not shown
Artery	669	327

FLEXOR HALLUCIS BREVIS

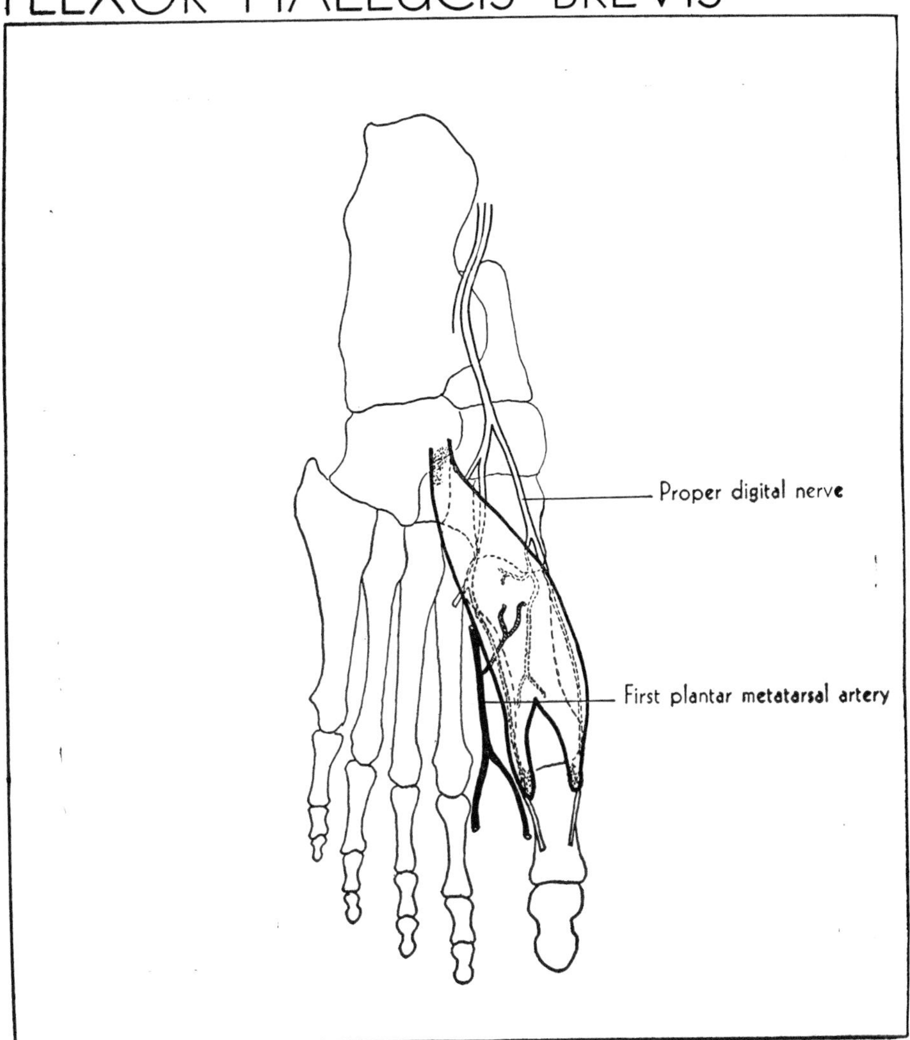
Proper digital nerve
First plantar metatarsal artery

ORIGIN: Medial part of plantar surface of cuboid bone, adjacent portion of lateral cuneiform bone, prolongation of tendon of tibialis posterior

INSERTION: Medial and lateral side of proximal phalanx of great toe

FUNCTION: Flexes great toe

NERVE: Proper digital nerve of great toe (1st plantar digital nerve) of medial plantar nerve

ARTERY: First plantar metatarsal (from junction of lateral and deep plantar aa.)

REFERENCES:	GRAY	GRANT'S ATLAS
Muscle	517	331
Nerve	517, 999	330
Artery	669	Not shown

ADDUCTOR HALLUCIS

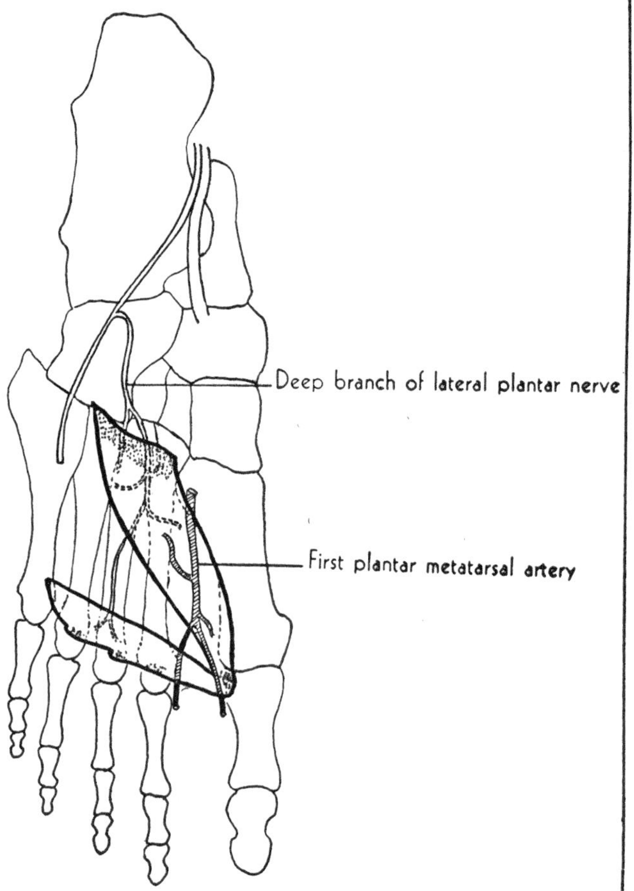

Deep branch of lateral plantar nerve

First plantar metatarsal artery

ORIGIN: Oblique head from bases of 2d, 3d, and 4th metatarsal bones, sheath of peroneus longus; transverse head from capsules of 2d, 3d, 4th, and 5th metatarsophalangeal ligaments, transverse ligament of sole

INSERTION: Lateral side of base of proximal phalanx of great toe

FUNCTION: Adducts great toe, assists in flexing it

NERVE: Deep branch of lateral plantar

ARTERY: First plantar metatarsal

REFERENCES:	GRAY	GRANT'S ATLAS
Muscle	518	330
Nerve	518, 999	330
Artery	669	Not shown

FLEXOR DIGITI MINIMI BREVIS

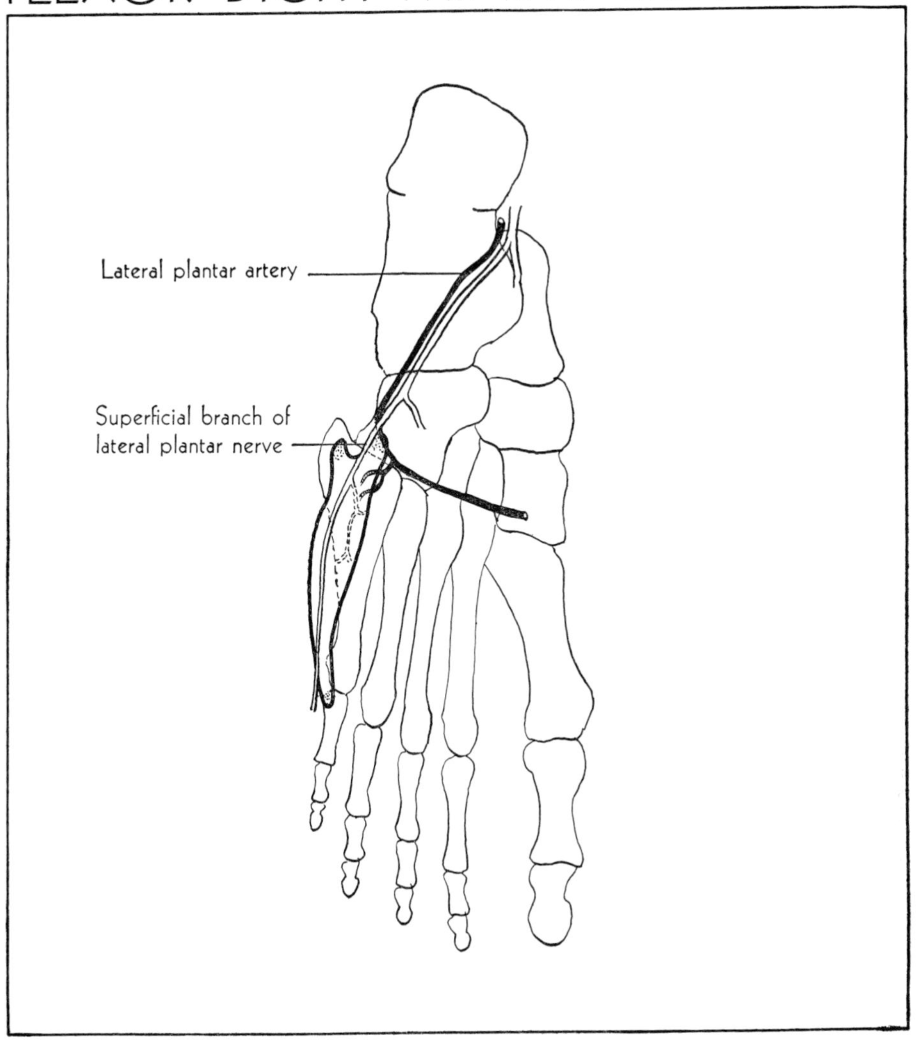

Lateral plantar artery
Superficial branch of lateral plantar nerve

ORIGIN:	**Sheath of peroneus longus, base of 5th metatarsal bone**
INSERTION:	**Lateral side of base of proximal phalanx of little toe**
FUNCTION:	**Flexes little toe**
NERVE:	**Superficial branch of lateral plantar**
ARTERY:	**Lateral plantar**

REFERENCES:	GRAY	GRANT'S ATLAS
Muscle	518	331
Nerve	518, 999	330
Artery	668	Not shown

INTEROSSEI DORSALES (Dorsal Interossei)

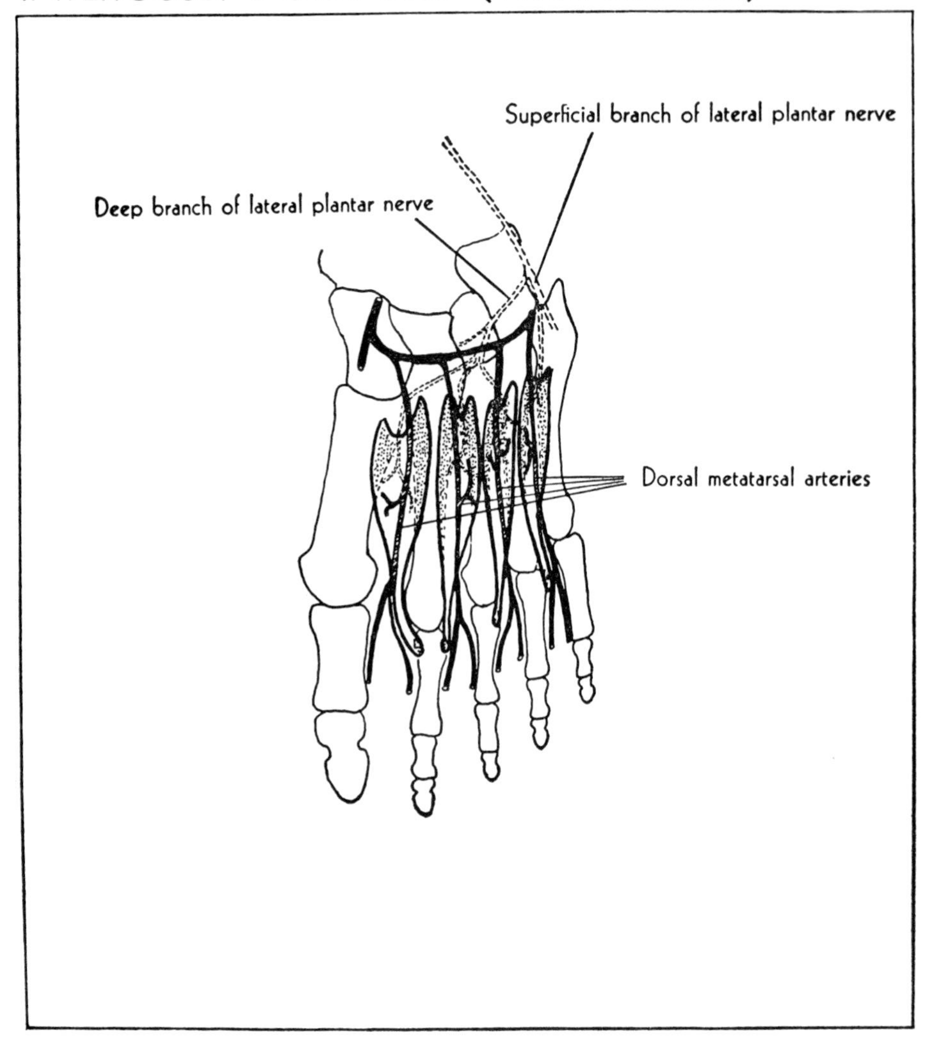
Superficial branch of lateral plantar nerve
Deep branch of lateral plantar nerve
Dorsal metatarsal arteries

ORIGIN: There are 4 dorsal interossei, each arising by 2 heads from adjacent sides of metatarsal bones

INSERTION: 1st into medial side of proximal phalanx of 2d toe, 2d into lateral side of proximal phalanx of 2d toe, 3d into lateral side of proximal phalanx of 3d toe, 4th into lateral side of proximal phalanx of 4th toe

FUNCTION: Abduct 2d, 3d, and 4th toes from axis of 2d toe, assist in flexing proximal phalanges and in extending middle and distal phalanges

NERVE: Superficial and deep branches of lateral plantar

ARTERY: Dorsal metatarsal

REFERENCES:	GRAY	GRANT'S ATLAS
Muscle	519	307, 331
Nerve	519, 999	330
Artery	666-667	306

INTEROSSEI PLANTARES (Plantar Interossei)

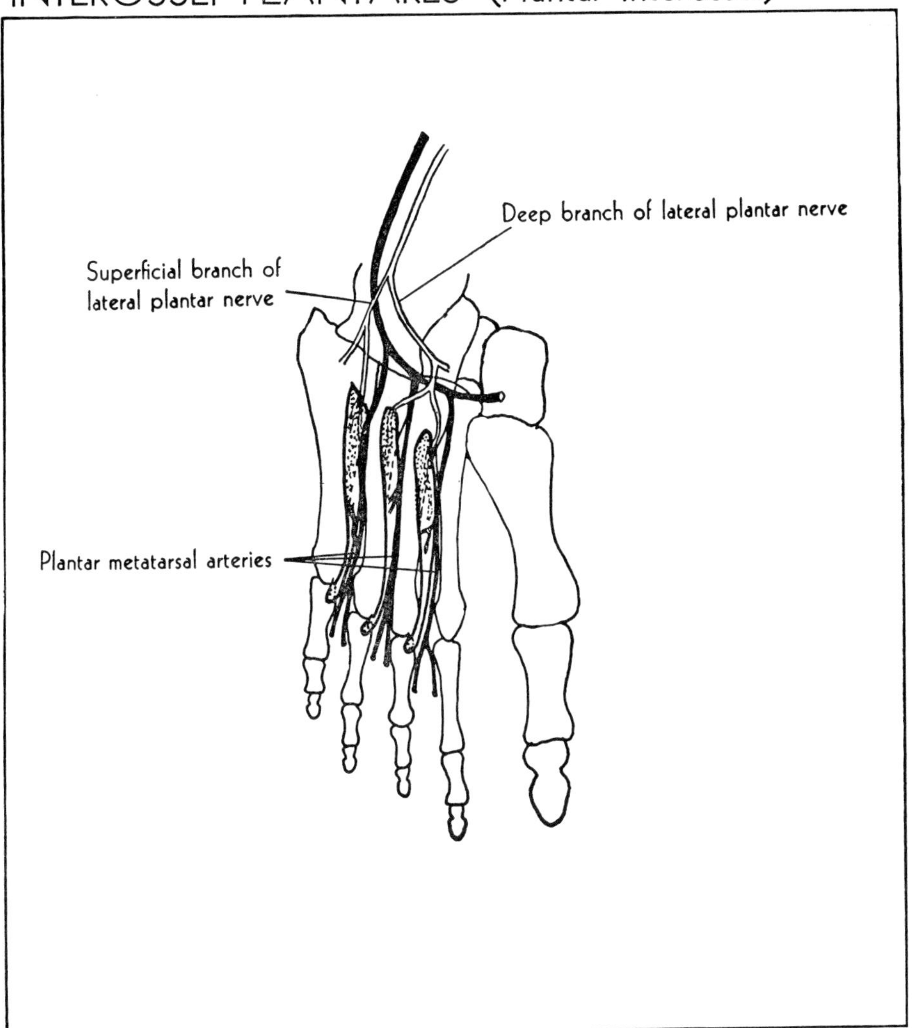
Deep branch of lateral plantar nerve
Superficial branch of
lateral plantar nerve
Plantar metatarsal arteries

ORIGIN: There are 3 plantar interossei, arising from bases and medial sides of 3d, 4th, and 5th metatarsal bones

INSERTION: Medial sides of bases of proximal phalanges of 3d, 4th, and 5th toes

FUNCTION: Adduct 3d, 4th, and 5th toes toward axis of 2d toe, assist in flexing proximal phalanges and in extending middle and distal phalanges

NERVE: Superficial and deep branches of lateral plantar

ARTERY: Plantar metatarsal

REFERENCES:	GRAY	GRANT'S ATLAS
Muscle	519	331
Nerve	519, 999	330
Artery	667, 669	Not shown

THE HUMAN SKELETON

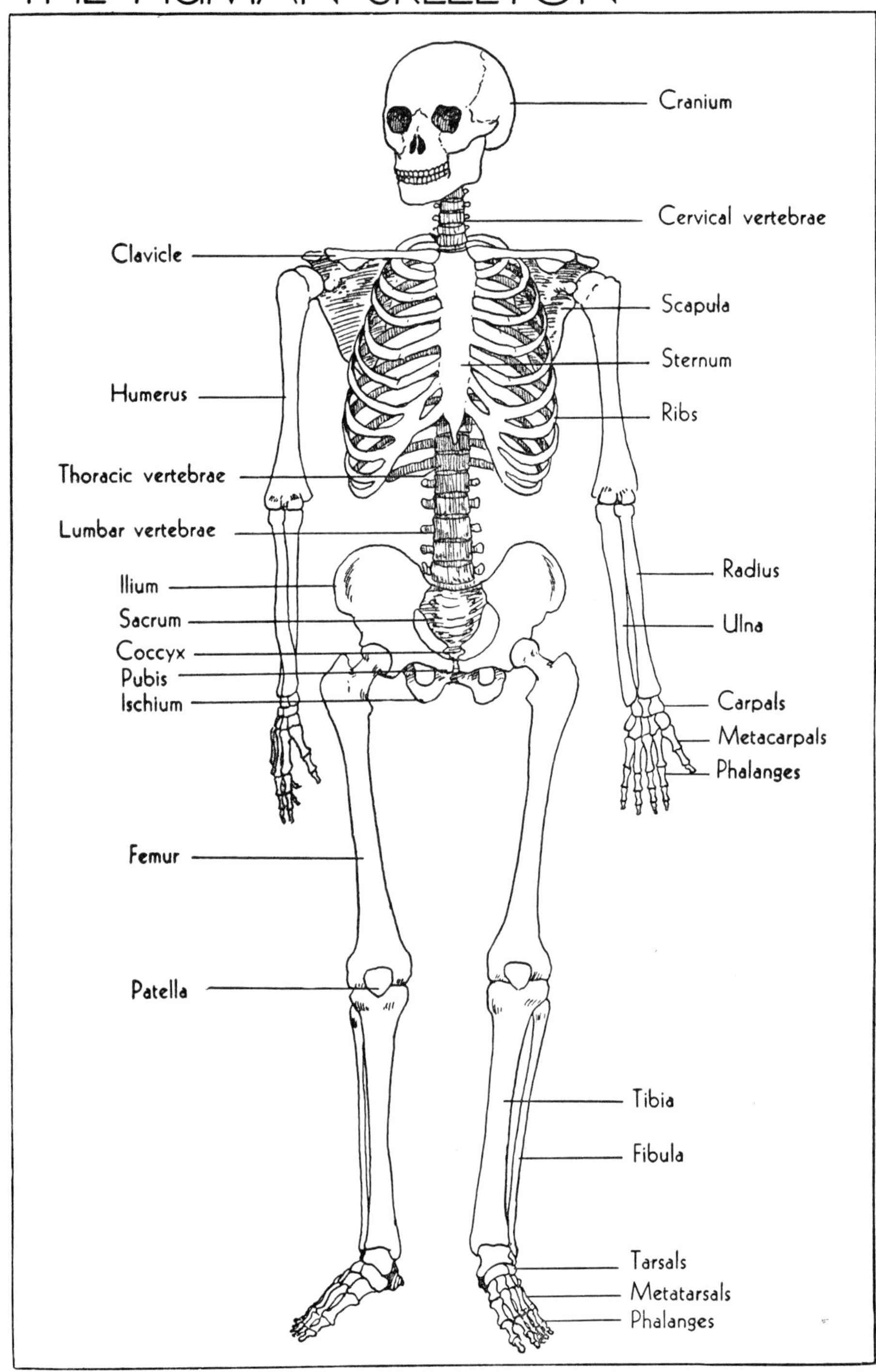
Cranium
Cervical vertebrae
Clavicle
Scapula
Sternum
Humerus
Ribs
Thoracic vertebrae
Lumbar vertebrae
Radius
Ilium
Sacrum
Ulna
Coccyx
Pubis
Ischium
Carpals
Metacarpals
Phalanges
Femur
Patella
Tibia
Fibula
Tarsals
Metatarsals
Phalanges

NERVES OF UPPER EXTREMITIES

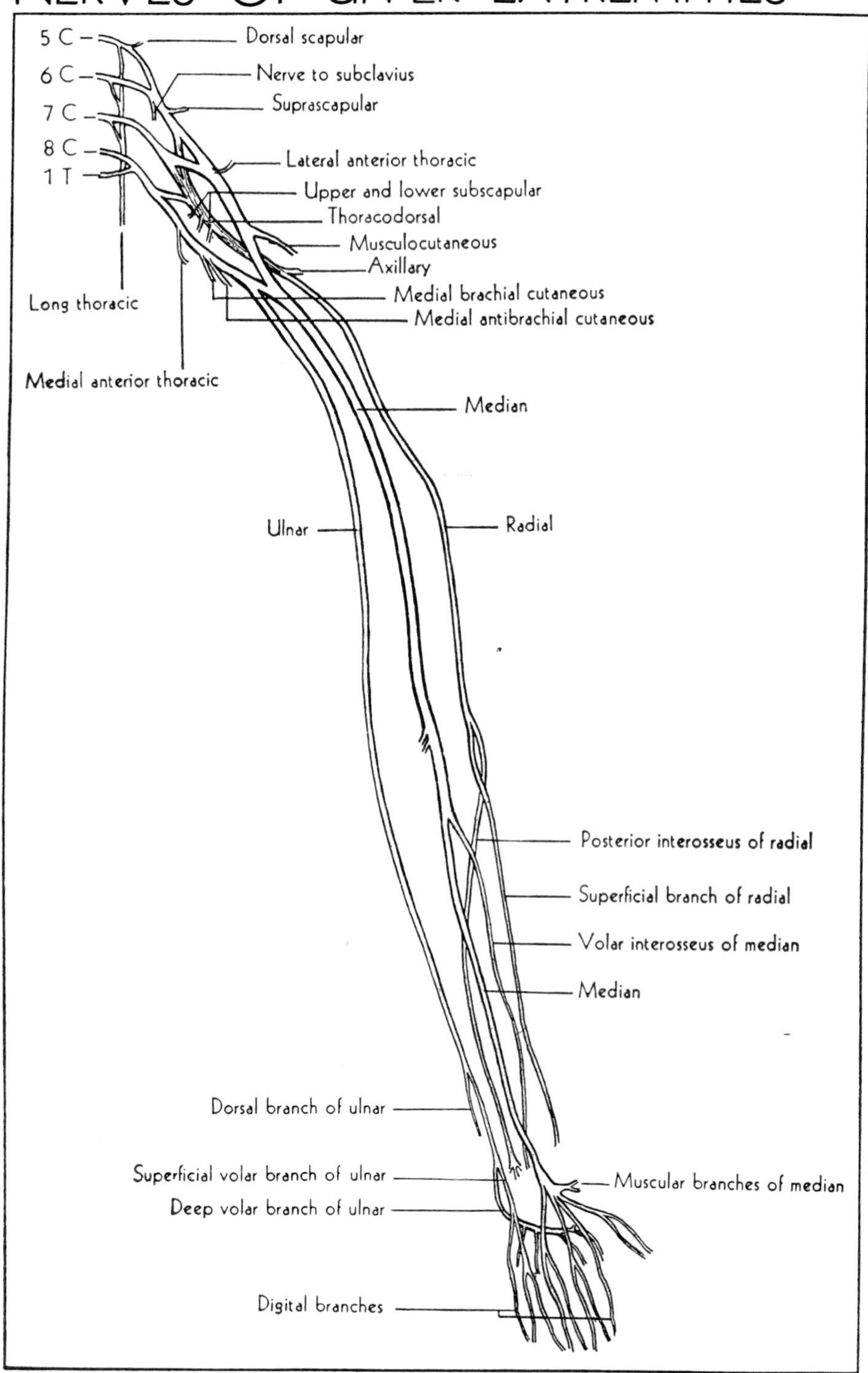

5 C
6 C
7 C
8 C
1 T
Dorsal scapular
Nerve to subclavius
Suprascapular
Lateral anterior thoracic
Upper and lower subscapular
Thoracodorsal
Musculocutaneous
Axillary
Medial brachial cutaneous
Medial antibrachial cutaneous
Long thoracic
Medial anterior thoracic
Median
Ulnar
Radial
Posterior interosseus of radial
Superficial branch of radial
Volar interosseus of median
Median
Dorsal branch of ulnar
Superficial volar branch of ulnar
Deep volar branch of ulnar
Muscular branches of median
Digital branches

ARTERIES OF UPPER EXTREMITIES

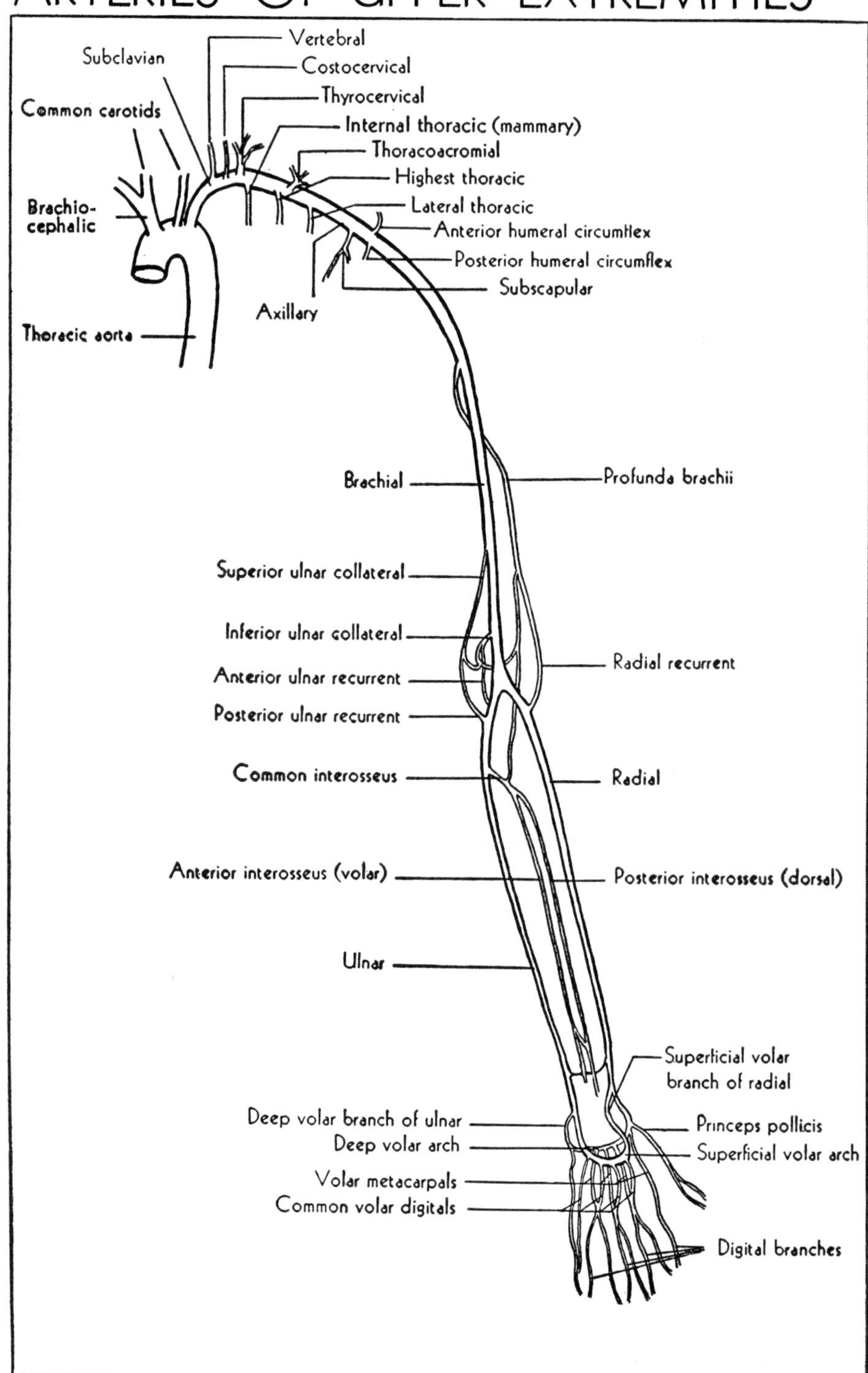
Vertebral
Subclavian
Costocervical
Thyrocervical
Common carotids
Internal thoracic (mammary)
Thoracoacromial
Highest thoracic
Brachio-
cephalic
Lateral thoracic
Anterior humeral circumflex
Posterior humeral circumflex
Subscapular
Axillary
Thoracic aorta
Brachial
Profunda brachii
Superior ulnar collateral
Inferior ulnar collateral
Radial recurrent
Anterior ulnar recurrent
Posterior ulnar recurrent
Common interosseus
Radial
Anterior interosseus (volar)
Posterior interosseus (dorsal)
Ulnar
Superficial volar
branch of radial
Deep volar branch of ulnar
Princeps pollicis
Deep volar arch
Superficial volar arch
Volar metacarpals
Common volar digitals
Digital branches

NERVES OF LOWER EXTREMITIES

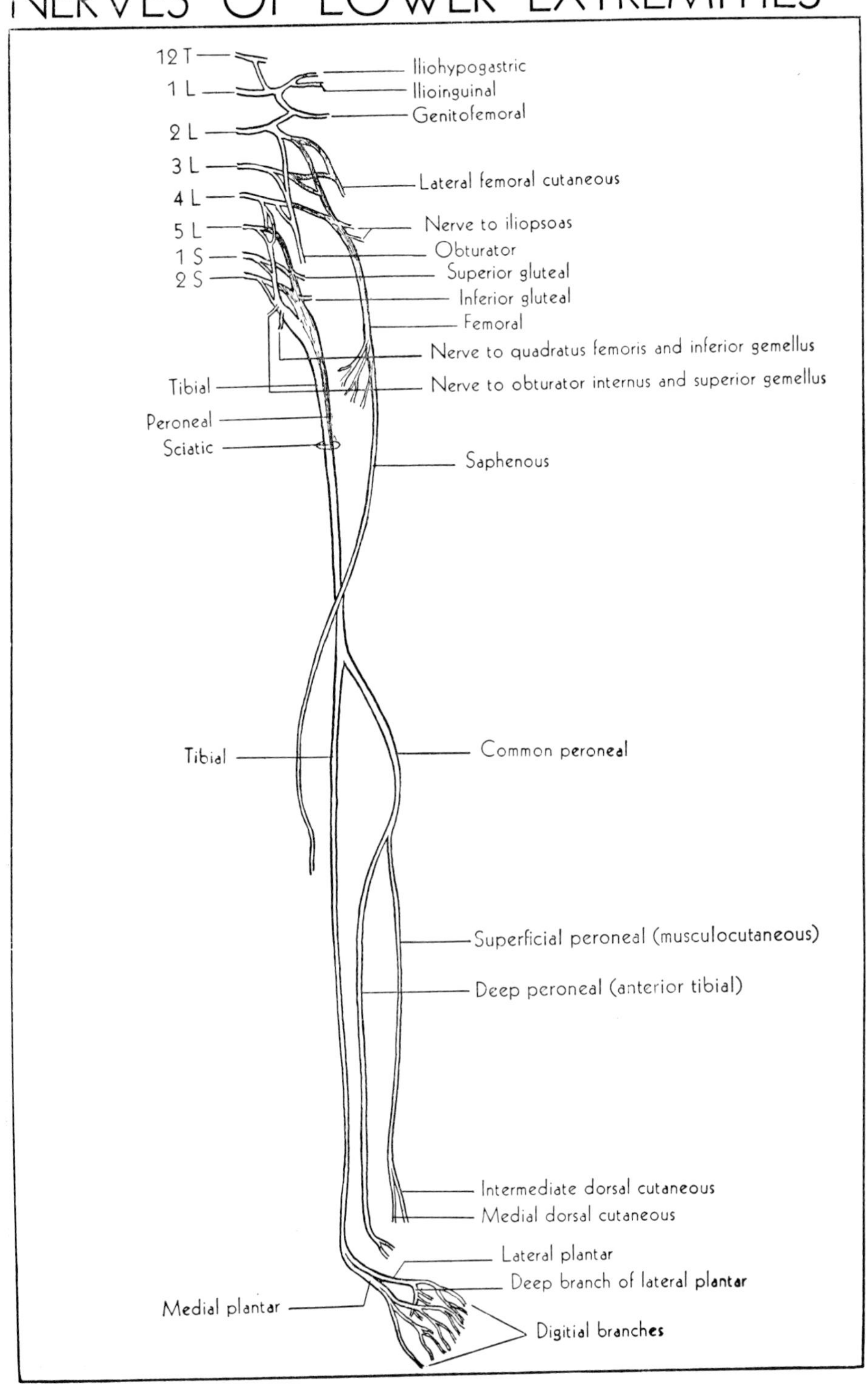
12 T
1 L
2 L
3 L
4 L
5 L
1 S
2 S
Iliohypogastric
Ilioinguinal
Genitofemoral
Lateral femoral cutaneous
Nerve to iliopsoas
Obturator
Superior gluteal
Inferior gluteal
Femoral
Nerve to quadratus femoris and inferior gemellus
Nerve to obturator internus and superior gemellus
Tibial
Peroneal
Sciatic
Saphenous
Tibial
Common peroneal
Superficial peroneal (musculocutaneous)
Deep peroneal (anterior tibial)
Intermediate dorsal cutaneous
Medial dorsal cutaneous
Lateral plantar
Deep branch of lateral plantar
Medial plantar
Digitial branches

ARTERIES OF LOWER EXTREMITIES

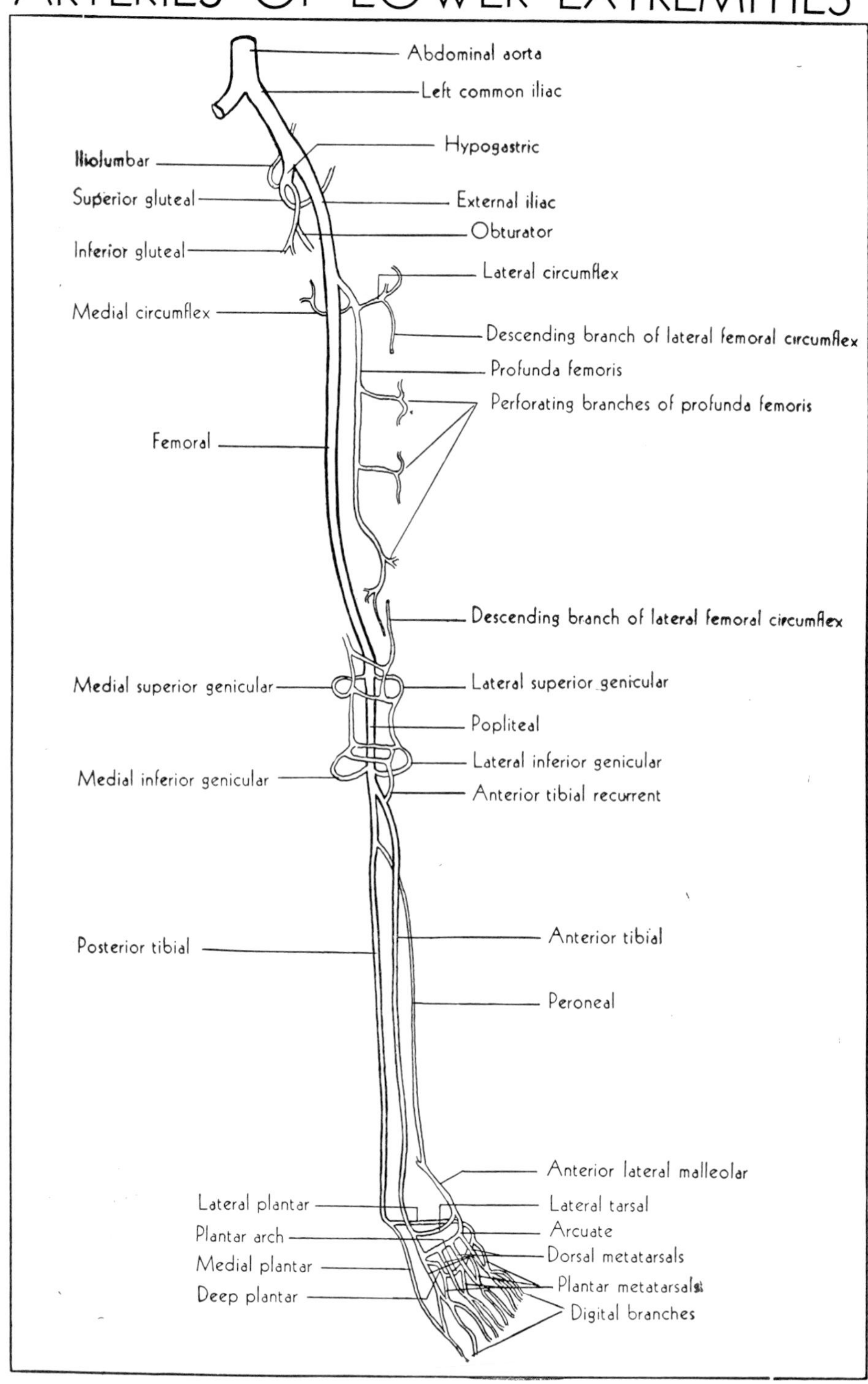
Abdominal aorta
Left common iliac
Hypogastric
Iliolumbar
Superior gluteal
External iliac
Obturator
Inferior gluteal
Lateral circumflex
Medial circumflex
Descending branch of lateral femoral circumflex
Profunda femoris
Perforating branches of profunda femoris
Femoral
Descending branch of lateral femoral circumflex
Medial superior genicular
Lateral superior genicular
Popliteal
Lateral inferior genicular
Medial inferior genicular
Anterior tibial recurrent
Posterior tibial
Anterior tibial
Peroneal
Anterior lateral malleolar
Lateral plantar
Lateral tarsal
Plantar arch
Arcuate
Medial plantar
Dorsal metatarsals
Plantar metatarsals
Deep plantar
Digital branches

MUSCLES OF RIGHT UPPER EXTREMITY—ANTERIOR VIEW
ORIGINS—SOLID INSERTIONS—STIPPLED

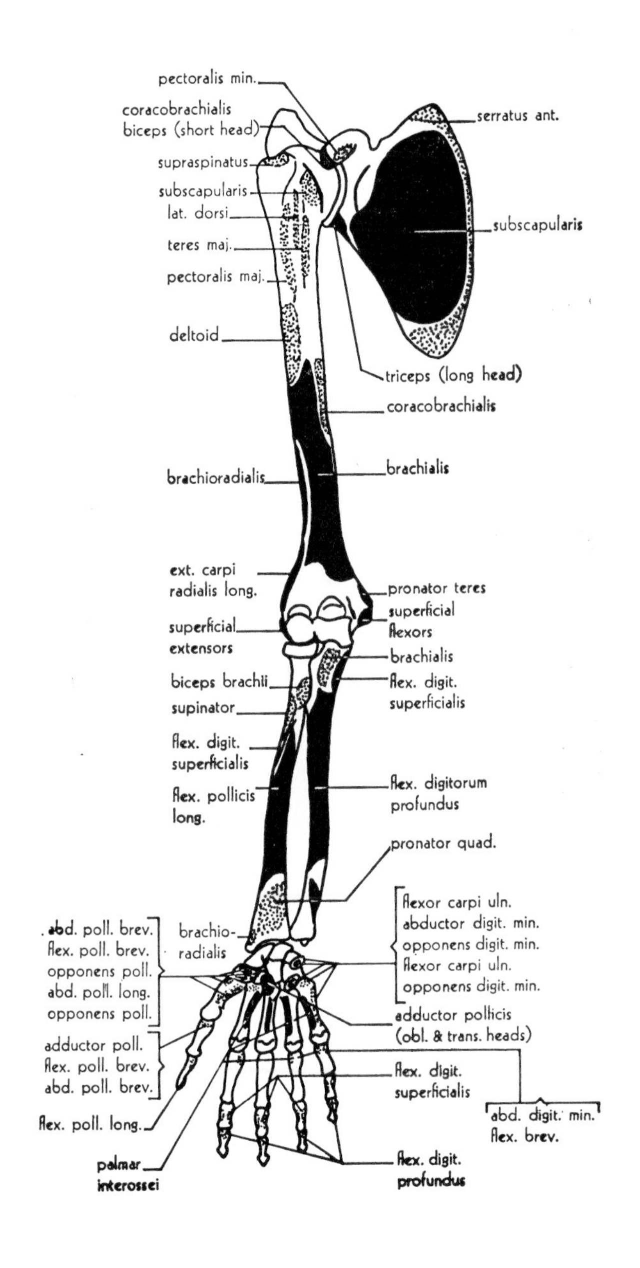
pectoralis min.
coracobrachialis
biceps (short head)
supraspinatus
subscapularis
lat. dorsi
teres maj.
pectoralis maj.
deltoid
serratus ant.
subscapularis
triceps (long head)
coracobrachialis
brachioradialis
brachialis
ext. carpi radialis long.
superficial extensors
biceps brachii
supinator
flex. digit. superficialis
flex. pollicis long.
pronator teres
superficial flexors
brachialis
flex. digit. superficialis
flex. digitorum profundus
pronator quad.
abd. poll. brev.
flex. poll. brev.
opponens poll.
abd. poll. long.
opponens poll.
brachio-radialis
flexor carpi uln.
abductor digit. min.
opponens digit. min.
flexor carpi uln.
opponens digit. min.
adductor pollicis (obl. & trans. heads)
adductor poll.
flex. poll. brev.
abd. poll. brev.
flex. digit. superficialis
abd. digit. min.
flex. brev.
flex. poll. long.
palmar interossei
flex. digit. profundus

MUSCLES OF RIGHT UPPER EXTREMITY—POSTERIOR VIEW
ORIGINS—SOLID INSERTIONS—STIPPLED

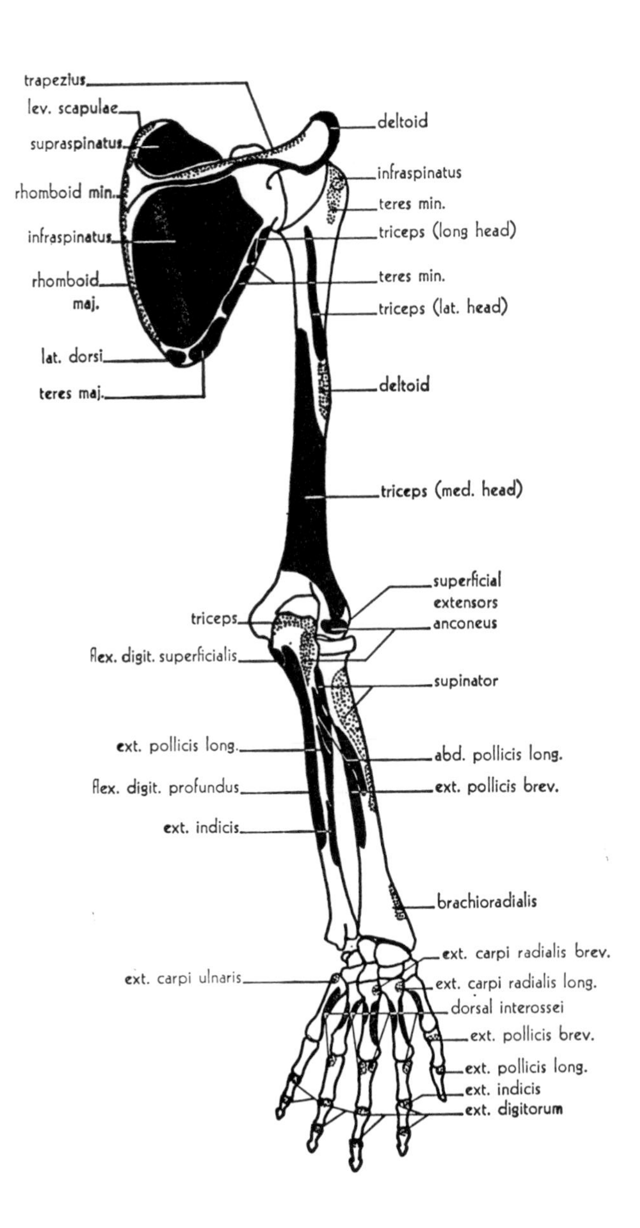

trapezius
lev. scapulae
supraspinatus
rhomboid min.
infraspinatus
rhomboid maj.
lat. dorsi
teres maj.
deltoid
infraspinatus
teres min.
triceps (long head)
teres min.
triceps (lat. head)
deltoid
triceps (med. head)
superficial extensors
anconeus
triceps
flex. digit. superficialis
supinator
ext. pollicis long.
abd. pollicis long.
flex. digit. profundus
ext. pollicis brev.
ext. indicis
brachioradialis
ext. carpi radialis brev.
ext. carpi ulnaris
ext. carpi radialis long.
dorsal interossei
ext. pollicis brev.
ext. pollicis long.
ext. indicis
ext. digitorum

MUSCLES OF RIGHT LOWER EXTREMITY
ANTERIOR VIEW
ORIGINS—SOLID; INSERTIONS—STIPPLED

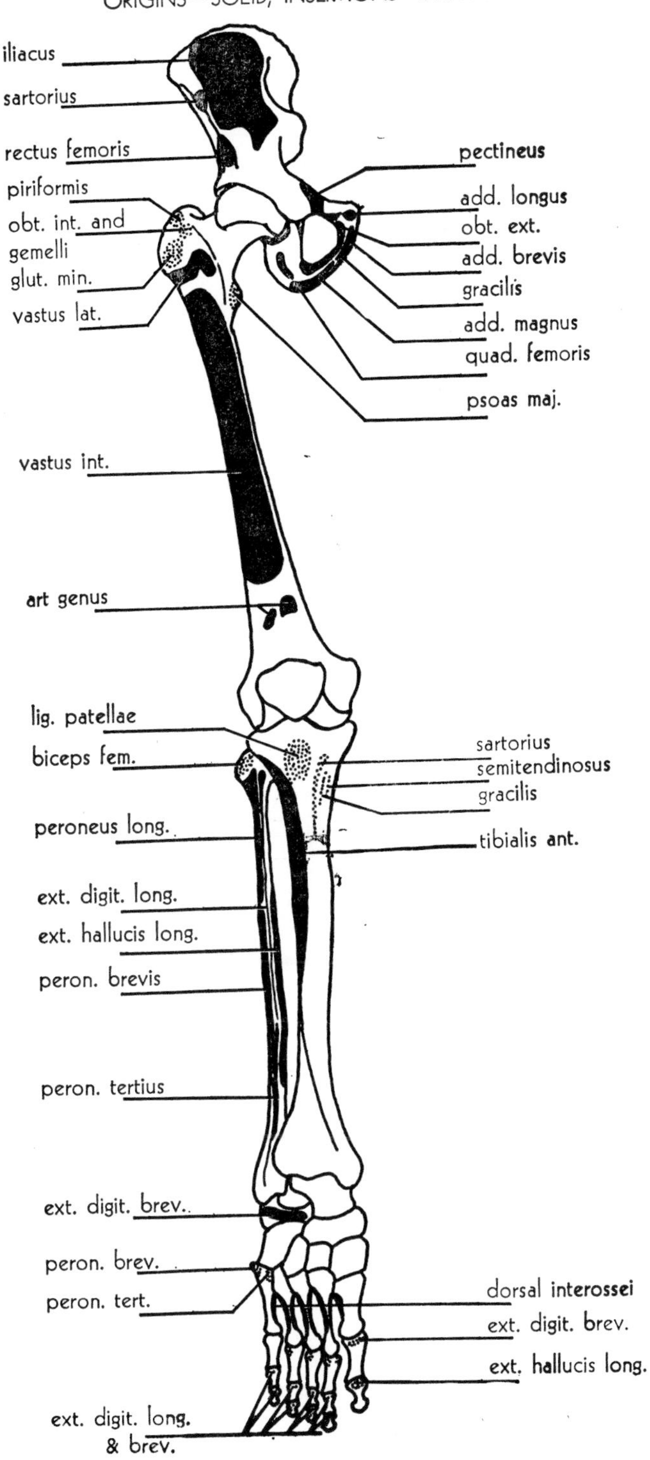
iliacus
sartorius
rectus femoris
piriformis
obt. int. and gemelli
glut. min.
vastus lat.
pectineus
add. longus
obt. ext.
add. brevis
gracilis
add. magnus
quad. femoris
psoas maj.
vastus int.
art genus
lig. patellae
biceps fem.
sartorius
semitendinosus
gracilis
peroneus long.
tibialis ant.
ext. digit. long.
ext. hallucis long.
peron. brevis
peron. tertius
ext. digit. brev.
peron. brev.
peron. tert.
dorsal interossei
ext. digit. brev.
ext. hallucis long.
ext. digit. long. & brev.

MUSCLES OF RIGHT LOWER EXTREMITY
POSTERIOR VIEW

ORIGINS—SOLID; INSERTIONS—STIPPLED

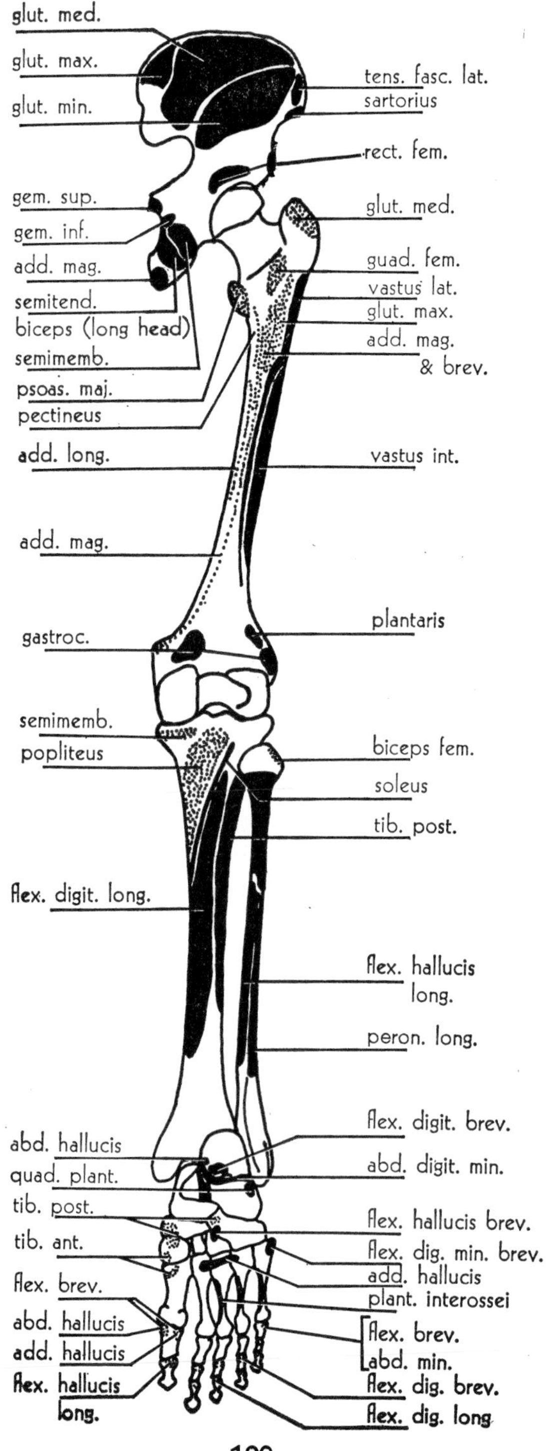

glut. med.
glut. max.
glut. min.
gem. sup.
gem. inf.
add. mag.
semitend.
biceps (long head)
semimemb.
psoas. maj.
pectineus
add. long.
add. mag.
gastroc.
semimemb.
popliteus
flex. digit. long.
abd. hallucis
quad. plant.
tib. post.
tib. ant.
flex. brev.
abd. hallucis
add. hallucis
flex. hallucis long.
tens. fasc. lat.
sartorius
rect. fem.
glut. med.
guad. fem.
vastus lat.
glut. max.
add. mag. & brev.
vastus int.
plantaris
biceps fem.
soleus
tib. post.
flex. hallucis long.
peron. long.
flex. digit. brev.
abd. digit. min.
flex. hallucis brev.
flex. dig. min. brev.
add. hallucis
plant. interossei
flex. brev.
abd. min.
flex. dig. brev.
flex. dig. long

INDEX

ARTERIES

INDEX—Continued

INDEX—Continued

MUSCLES—Continued

NERVES

NERVES—Continued